INTERNATIONAL SERIES OF MONOGRAPHS IN
EXPERIMENTAL PSYCHOLOGY

GENERAL EDITOR: H. J. EYSENCK

VOLUME 16

ASPECTS OF MOTION PERCEPTION

OTHER TITLES IN THE SERIES IN EXPERIMENTAL PSYCHOLOGY

ASPECTS OF
MOTION PERCEPTION

BY

PAUL A. KOLERS

PERGAMON PRESS

OXFORD · NEW YORK · TORONTO
SYDNEY · BRAUNSCHWEIG

Pergamon Press Ltd., Headington Hill Hall, Oxford
Pergamon Press Inc., Maxwell House, Fairview Park, Elmsford,
New York 10523
Pergamon of Canada Ltd., 207 Queen's Quay West, Toronto 1
Pergamon Press (Aust.) Pty. Ltd., 19a Boundary Street,
Rushcutters Bay, N.S.W. 2011, Australia
Vieweg & Sohn GmbH, Burgplatz 1, Braunschweig

First edition 1972
Library of Congress Catalog Card No. 73–188746

Printed in Great Britain by A. Wheaton & Co., Exeter

08 016843 4

FOR

JULES V. COLEMAN

AND

GEORGE A. MILLER

WHO MADE LIGHT OF DARKNESS

Human vision is such an enormously rich complex of experiences, and human beings are so diversified in habits and interest, that no two of us value our eyes for quite the same set of reasons. If asked what aspect of vision means the most to them, a watchmaker may answer "acuity," a night flier, "sensitivity," and an artist, "color." But to the animals which invented the vertebrate eye, and hold the patents on most of the features of the human model, the visual registration of **movement** *was of the greatest importance.*

G. L. WALLS (1963, p. 342)

CONTENTS

FOREWORD

In 1964 I published a paper that pointed to certain differences in the way the visual system implemented the perception of motion derived from physically moving objects, and illusory or apparent motion derived from flashing lights. The paper came to the notice of H. J. Eysenck, editor of the International Monograph Series in Experimental Psychology that is published by Pergamon Press. I accepted his invitation to prepare a monograph on the subject of apparent motion, but between that time and 1969 my attention was occupied by other things. At the latter date, at Bell Telephone Laboratories, I undertook the study of some aspects of motion and form perception, particularly as they relate to the quality and fidelity of electronically transmitted images. This monograph grew out of that study.

A number of problems confront the engineer interested in transmitting pictures electronically, because pictures tend to be quite detailed but the transmission medium is both limited and expensive. Hence a number of engineers actively explore the topic of bandwidth compression, seeking a means by which maximum amounts of visual information can be transmitted at minimum cost. It seemed to me that a contribution to this topic that a psychologist might make would be to study the kinds of information that people actually use in their perception of continuing sequences of briefly presented pictures, such as characterize television transmission of real scenes. Not everything in a picture is perceived, nor do people typically have to look at great amounts of detail in order to come away with an adequate representation of what they are being shown. The reason is that much of our perceiving is anticipatory and inferential; it is based on an active sampling of clues rather than on passive reception of details. What is not known, however, is whether the sampling is haphazard or whether it is rule-governed; whether, that is to say, any aspect of any object is as good a clue to its identity and is as likely to be sampled as any other. The intuitive answer to this uncertainty is a negative one; and in this regard concepts of salient or distinguishing features that characterize objects and facilitate their identification have come to the fore in recent years. Hence we may ask, what aspects of pictures best aid their recognition; what features most readily aid in distinguishing them? What are, in fact, the salient or distinctive features of pictures? Can one formalize these characteristics?

Questions of this kind motivated some of the research I shall report later in this monograph. The direction followed in seeking an answer was the classical one of dealing with simple, stripped down instances of stimuli. Although the pictures most likely to be transmitted in an electronic system are of people in movement, we began with simpler instances; in fact, with such classically simple ones as plane geometric figures. Moreover, we studied their perceptibility not when they were presented one at a time, measured for their specific geometry and luminance, but when they were presented in brief sequences. Therefore we were immediately dealing with illusory motion of the configurations, for it is a characteristic of the visual system that it perceives brief, properly sequenced presentations as objects undergoing spatial translation. It is just this characteristic, in fact, that the motion-picture projector and the television screen take advantage of, albeit in substantially different ways, in triggering those perceptions of objects in motion that we do see with those media. As we shall see below, the technological successes with those media do not depend upon a correct understanding of their effects on the nervous system; the technology works, but no currently available account of the perceptibility of forms in motion is correct. Theory lags practice rather extensively in this domain.

One outcome of our research was the discovery that the visual system actively constructs, implements, and fills in features of objects derived from a sampling of the physical presentation. The visual system does not merely record what has been presented to it (if it records that at all); it actively creates its own pictorial reality. Moreover, the mechanisms underlying this creation seem to be substantially different from the mechanisms utilized in feature analysis. What role, if any, that feature analysis plays in normal perception is not entirely clear.

These facts emerged from a study I carried out with James R. Pomerantz in the fall and winter of 1969 and published in the *Journal of Experimental Psychology* in 1971. Intrigued and puzzled by them, but under a deadline created by other forces, we followed through with this investigation during the early spring of 1970. Reminded by Pergamon Press of my committment to do a monograph on apparent motion, I thought that these new investigations might be embedded in the context of this old problem; for the fact is that people have been looking at motion pictures for more than seventy years, and at television screens for more than twenty-five, but the mechanisms governing these perceptions are still unknown.

Regrettably, our work does not answer all of the questions it set out to answer. Indeed, we were blocked for some considerable time, for in our thinking we continued to seek positive answers to certain questions whereas it becomes increasingly likely that some of them cannot be answered positively. Thus for conceptual and administrative reasons it was not possible to round out this work, and it is incomplete. I present it at least for the questions it raises if not for the answers it provides.

On and off between 1964 and 1968 I discussed various aspects of the perception of illusory motion at Project Zero at the Harvard Graduate School of Education. I think it was in response to those beneficent challenges that some concern with pictorial notation crept into my own thinking, although whatever distortions it may have there are not to be laid at the door of Project Zero or its director, Professor Nelson Goodman. It was during that period also that I carried out some preliminary experiments on form and motion perception at the Research Laboratory of Electronics, in connection with a course on perception and cognition I taught in the Department of Electrical Engineering at Massachusetts Institute of Technology. Moreover, as my own command of German is poor, I am indebted to several friends and former colleagues for aid with various aspects of the older literature. In particular I thank the former Misses Katharine Gilbert and Jean Mechlowitz, and Dr. D. C. Milne in this connection. I also thank Drs. David N. Perkins, Ronald Cohen and Ivan Bodis-Wollner for their comments on the manuscript. My greatest obligation, however, is to James R. Pomerantz for his conscientious, creative, and energetic contributions to our short-lived collegiality. I am not sure that I would have had the patience or subtlety to pursue these elusive realities alone; indeed he always shared and sometimes even led in the formulation of our experiments. And of course I shall forever remember my stay at Bell Telephone Laboratories, and here record my thanks to that institution for enabling me to carry out this research.

Toronto, Ontario

ACKNOWLEDGEMENT

I thank the authors and the American Psychological Association for permission to reproduce the following figures: 3.6, 4.3, 4.4, 4.5, 6.2.

CHAPTER 1

THE BACKGROUND

ABOUT 2500 years ago the mathematician and philosopher Zeno of Elea invented four puzzles regarding distance and motion which were, in Bertrand Russell's words, "immeasurably subtle and profound". The most famous of them, that the swift Achilles can never overtake a tortoise that has started a race before him, and indeed all of the other three, have finally given way before mathematical analysis (Russell, 1938, pp. 347–354; Grünbaum, 1967), but like a ghost in the machine, they still haunt proposals about the mechanisms of perception.

One interpretation of Zeno is that the perception of motion is based not on current sensory information, but on memory for position and time; hence on comparison, guess, or inference. This interpretation alleges that what our visual system actually detects are objects in different locations at different times; noting the disparity, we create a sense of motion to resolve it. Perception of objects, memory of their position, and delusion are therefore the main components, according to this theory, of our perception of motion.

Until the last quarter of the nineteenth century there was little evidence that refuted this view directly. Then, in 1875, the physiologist Sigmund Exner showed that when things are arranged properly, two brief but stationary flashes are seen as a single object in motion. The timing of events is so rapid moreover, and the perception of motion so immediate, that memory cannot be its source. Working at a time when the task of psychology was taken to be the identification of the sensory elements of perception, Exner argued that motion is not an inferred attribute of objects perceived in different places, but a basic element in the mind's armamentarium.

Aside from its philosophical implications, the import for psychology of Exner's study lies in its systematic use of what even at that time

1

was a well-known laboratory curiosity, the phenomenon of apparent motion. To create the illusion for his experiment he exposed two spatially separated electrical sparks sequentially; when the timing was appropriate, his observers saw a single flash move smoothly across the empty space between them. Exner's interest lay less in the characteristics of this perception of motion, however, than in establishing motion as a basic sensation; thus, except for an occasional paper, it was not until 35 years later, when Max Wertheimer began his systematic work, that this old phenomenon of illusory motion was subjected to intensive and detailed study. It is worth speculating about the delay.

One possibility is technological. Creating the phenomenon for controlled study requires the use of special apparatus; electric sparks are not very good sources of illumination in visual experiments that manipulate time as a variable. Perhaps this is one reason that the phenomenon was not explored extensively; but if it is, it is not a strong reason, for many mechanical contrivances were known that yield vivid perceptions of motion (Boring, 1942). A more convincing reason for the neglect of the phenomenon lies with the intellectual Zeitgeist of the nineteenth century. The notion then current of how the mind works is in some respects different from our own. Many scientists then thought that the visual system analyzed stimuli to extract the important elements which, following Helmholtz, were called "sensations", and the mind combined these into "perceptions". Moreover, the sensations were long thought to be faithful to the reality of the physical world, representing it truly. Therefore many scientists of that period believed that perception stood in a direct relation to the objects inducing the stimulation, so that our perceptual experiences were faithful copies of the physical world. Apparent motion, however, like many other perceptual experiences, is an illusion; it is a believable perception of motion that originates in the sequenced flashing of physically stationary objects. The study of illusions did not have much respectability in an intellectual environment that emphasized veridicality of perception. As late as the end of the century the distinguished German psychologist Oswald Külpe (1893), who actually was Wertheimer's teacher, wrote of illusions that they are (in Titchener's translation) "subjective perversions of the contents of objective perception".[1] By "objective perception" he was assuming the

[1] "subjective Veränderungen an dem objectiv Wahrnehmbaren" (p. 184).

correctness of just that mechanistic philosophy that argued for a one-to-one correspondence between physical stimulation and psychological experience. The phenomenon of apparent motion is a dramatic violation of that assumed correspondence.

The perception of motion from the sequential illumination of discrete stationary objects is so compelling perceptually and so significant theoretically that hundreds of papers have been published on its various aspects in the past 60 years. I have already alluded to its significance for Exner as a demonstration that motion is a direct rather than a derived perceptual experience. In Wertheimer's hands the phenomenon took on even deeper significance; indeed, his paper of 1912 created the foundation for one of the most vigorous branches of experimental psychology in the twentieth century—Gestalt Psychology.

Consider that Wertheimer was working at a time when the Wundtian and Titchnerian form of elementarism was in full flower; this point of view derived from British empiricism in general and from the associationism of the Mills in particular (Heidbreder, 1933). The view held that our perceptual experiences are complex events analogous to chemical compounds, whose content represents the proper intermixture of more simple elements. This mental chemistry, as it was called, took as its major task identifying the basic sensory elements and the rules by which they combined. The rules of combination were thought to be known; these were the laws of association of qualities and attributes ("ideas") that the British empiricists from the seventeenth century on had worked out. The machinery of combination known, the task remaining was to identify the elements it worked on. Then, it was thought, we would have a full accounting of mental contents, the perceptions and thoughts of everyday experience.

Wertheimer, a deeply imaginative man, questioned this interpretation of perceptual processes. Perception cannot be so elementaristic, he argued; for example, what are the elements of a rhythm and how, as von Ehrenfels had pointed out earlier, could such an elementarism explain the perception of a melody despite a transposition of key that left no notes in the two forms identical? Rather, Wertheimer argued, the nervous system is a device that organizes and structures its sensory inputs; it does more than convey sensations as so many electrical signals through so many wires. The introspectionist psychologists, in insisting

upon analysis and decomposition, seemed to ignore the basic characteristic of perception, that it is of whole objects segregated in space and
time. The perception of motion from two stationary flashing lights is a
powerful example of the organizing characteristics of the visual nervous
system.

The illusion of motion can be created in many ways; in fact, there
are a number of illusions of motion. The one that is most familiar to
psychologists and that was exploited by Exner (1875) and studied by
Wertheimer (1912) has many names, among them phi motion, beta
motion and optimal motion. With present-day electronic technology,
a very easy way to produce this phenomenon is to illuminate two
small spatially separated gas-discharge lamps in sequence for brief
controlled durations. The lamps may be viewed directly, against a dark
background, or they may be used to illuminate drawings or other figures.
A multichambered tachistoscope that illuminates different objects
successively is a useful device for the purpose, but a computer generating
displays on a cathode-ray tube (CRT), or a motion picture projector
can also be used. So too can a stroboscope that intermittently illuminates
a physically moving object. Each method introduces slight variations
in the data that are obtained, for each method creates its own artifacts,
but all share the ability to create a perception of motion from stationary
objects illuminated in sequence. Wertheimer (1912), who was the first
to recognize many of the psychological and philosophical implications
of the phenomenon, did his own experiments mainly with a tachistoscope, and a borrowed one at that.

With great ingenuity Wertheimer analyzed many constituents of the
phenomenon. His experiments established the paradigm for most of the
later research, and many later papers worked out in detail conditions
and effects that he first described. He was, however, not always correct
in his conclusions.

One might think that the great amount of work expended on the
perception of illusory motion would have left little that was new to add.
In point of fact, however, the phenomenon has never been satisfac-
torily explained. Although several lines of theory have been advanced
and explored, none fully accounts for the facts in a detailed way. Indeed,
in the hands of the Gestalt psychologists the phenomenon was principally a vehicle for advancing their philosophy; in life, its major occur-

rence is on the motion picture screen, and on advertising signs that when properly flashed seem to move. Does so behaviorally trivial a phenomenon have any significance in the modern world? Is the phenomenon still worth investigating? And if so, why?

My answer is obviously affirmative on all counts. The basis of the affirmation is the recurring similarity of the data obtained in the study of illusory motion to data obtained in other studies of visual functions. Although illusory motion itself occurs infrequently in our daily lives, its occurrence seems to depend upon the action of mechanisms that are centrally involved in all perceptual experience. Man depends heavily upon his visual apparatus as the source of much of his information. It should be obvious that the better we understand how his visual apparatus works, the better we are able to evaluate and understand other related forms of information processing. The study of man's characteristics as an information-processing organism is the object of a large segment of contemporary psychology.

A related reason for the affirmation is that the phenomenon of apparent motion speaks to the broad issue of the means by which man represents to himself the characteristics of the world outside his own skin. Earlier philosophers and psychologists believed that the representation was quite straightforward and faithful to physical reality. We now know that this view is too simple; we substitute for it the view that man samples his environment, and on the basis of samples constructs a representation that may be more or less consensually verifiable. Wertheimer used the phenomenon of apparent motion to demonstrate some of the ways the nervous system organized stimulation it receives; its further analysis reveals much about the interplay in perception of environmentally-supplied and self-supplied information.

In the early chapters of this book I shall summarize some of the major findings about illusory and veridical motion and the theories they have generated. The central section of the book describes a number of experiments J. R. Pomerantz and I carried out to evaluate some of the theories. Regrettably, all the theories are invalidated. In later sections it will be shown that a two-component model of motion perception is required, but that even this is not fully adequate to the facts. In the final chapter I shall discuss some characteristics of the status of perceptual experiences, and show there that the visual system seems to contain two relatively independent means of representing experience pictorially.[2]

Before proceeding, however, it is necessary to distinguish between illusory and veridical perceptions on certain grounds, if only to explain the use of the words. We mean by "veridical perception" that information we acquire through one sensory channel is consistent with information we acquire through another; or that what we expect on the basis of acquired information can be tested by action; or, sometimes, that what we perceive is not entirely idiosyncratic but can be perceived by other people as well. Consensual verification is one test of veridicality. We refer by "illusory perception" to an experience that can be taken as an instance of something else. Information obtained by one means allows us to make predictions about consequences or concomitants of our experience; when these are not borne out, we define the perception as illusory. An object that seems to be moving should collide with objects in its path or should pass reference markers in a particular way. Failure to do so reveals that the object seen only mimics some of the features expected of the veridical perception, hence is illusory.

Illusions are illusory just because they share some notable features with events they mimic. Lacking certain crucial features, they do not make a perfect "fit" with our implicit theory of the world. It should be obvious, therefore, that appearance alone does not usually distinguish between veridical and illusory perceptions, for the illusion consists just in the fact that there is a compelling similarity of appearance between them. Hence careful analytical tests are often required to establish whether a given experience is illusory or not. Judgments of "reality" and truth cannot be reliably based on appearance only; they require confirmation from tests outside the experience itself, for an event cannot reliably be used to evaluate itself. By "veridical", therefore, we mean consensually verifiable to the limits of possible tests; and by "illusory" we mean looking like the veridical but not similarly verifiable. These are clumsy distinctions at best, but are necessary to keep in mind hereafter.

[2] Throughout, I use the words object, contour, shape, and figure in approximately synonymous sense to refer to what is perceived. Speaking strictly, of course, figure, shape, color, motion, and the like are all aspects or properties of the perceived object. In the same loose way I use "pictorial experience" as synonymous with visual perception.

CHAPTER 2

WERTHEIMER'S CONTRIBUTION

WERTHEIMER's paper of 1912 reported an astonishingly large number of effects. His ingenuity in testing the limits of the phenomenon of apparent motion and of devising critical tests for hypotheses was very great indeed. The paper not only launched the movement that became Gestalt Psychology, it also established the paradigm that is still followed in many studies of apparent motion. The paper is systematic, ingenious, and interesting, one of the great ones in the experimental psychology of visual perception. It is fitting therefore to begin by summarizing some of the ideas that exercised Wertheimer, and the demonstrations he made in response to them.

The argument that Wertheimer inherited from Exner (1875) was that movement is a sensation in its own right, one of the basic constituents of perception, and not a derived or computed one. Exner's affirmation of this idea was based on one of the first serious investigations utilizing apparent motion. (The idea itself was supported by several other contemporary investigators, whose work Boring (1942) has summarized.) Exner wished to measure the threshold for temporal succession. For the purpose he exposed two spatially separated electric sparks in sequence, varying the interval between them. He found that at temporal separations of about 45 msec the observer could regularly report the physical order of the flashes, but if the distance separating them was not too great, the observer perceived not only their order but motion between them. Exner then went on to measure the lower threshold for this effect, and found that observers attributed a direction to the perceived movement when the two flashes were separated by as little as 14 msec; and this occurred even though at these temporal intervals but at wider spatial separations their order could not be reported correctly.[1]

Exner showed, therefore, that three distinct perceptual events can be established with brief properly sequenced flashes. When the temporal separation is 10 msec or so the two flashes appear simultaneous; at somewhat longer separations they appear as a single object in motion; and at still longer separations they reappear as two flashes but in succession. The perception of motion, he found, occurred at shorter temporal intervals than were required for a perception of sequence or order. Therefore, Exner argued, memory of position and perception of order cannot be the basis of the perception of motion. Having made this telling point, Exner went on to other things, and the phenomenon itself was little explored for 35 years.

Between Exner's work and Wertheimer's, which began in 1910, the motion picture camera and motion picture projector were invented. Here were devices that yielded complex perceptions of motion indeed. The projector works by illuminating successive frames of film at a fixed rate, a shutter interrupting the beam of light while the film moves. The sequence projected on the screen is always therefore physically stationary, for the motion of the film is not itself projected. (If it is, by eliminating the shutter, one sees only smears of contours and shades.) According to one legend it was in contemplating the physiological and psychological aspects of the motion picture—the perception of motion from discrete presentations—that Wertheimer became interested in the phenomenon of apparent motion.

For his experiments, he simplified the situation drastically. Rather than with complex scenes, he began his investigation with a limiting case, two horizontal lines, or a horizontal and an oblique line, exposed in sequence. He readily found Exner's three stages when he varied the temporal interval between the lines. When the interval is very short, the two lines appear simultaneous and, when one is oblique, as forming an angle; when the interval is long enough they are seen in succession; and in an intermediate range they are seen in optimal movement: one

---

[1] The values are in rather good agreement with those found by recent investigators using more refined apparatus. See Hirsh and Sherrick (1961) for the perception of order, and Thorson, Lange, and Biederman-Thorson (1969) for direction of motion.

line appears to move through the physically empty space from the first position to the second. Wertheimer set himself several questions to answer.

The main question was whether he could clarify the emergence of optimal apparent motion—the perception of an object moving through the phyiscally empty space between the origin and terminus—from simultaneity at one end and succession at the other. He proposed to study this by varying the temporal interval between the flashes in small steps. Another question concerned the epistemological and perceptual status of the illusory object. He studied this in part by interposing other stimuli into the regions through which the illusory object moved. A third question was the way in which set and attitude affected the illusion. This he studied by varying instructions to the observer, to affect the location of eye fixations and what Wertheimer called the posture or attitude of attention. And finally he studied some aftereffects, but with a radically different kind of display from that used in the foregoing, a rotating arithmetic spiral. Except for the last tests, the main variables were the spatial and temporal characteristics of the flashes, whose duration and distance apart were manipulated. He also varied their configuration, but not in the same systematic way.

Wertheimer's experiments revealed quickly that the three stages Exner had described could be supplemented to include partial motions, and phi or objectless motion. The partial motions were characterized by a seeming movement of the first flash part way across the screen, where it disappeared, and the appearance of the second flash displaced from its true location but seeming to move toward the terminus (see Fig. 2.1). Phi motion, which was given the greatest theoretical attention, was described as a sense of motion without a concomitant perception of moving objects; its closest analog in physical terms would be an object that appears at one location, disappears at another, and moves so rapidly between them that only its motion but not details of its shape or identity can be made out.

A word on nomenclature is needed before proceeding. The term "phi phenomenon" or "phi movement" is sometimes used generically for illusory motion. This is misleading usage, for phi motion correctly refers only to global "figureless" or "objectless" apparent motion, analogous to the very rapid passage of a real object across the field

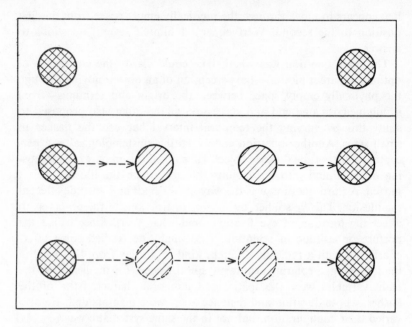

FIG. 2.1. At rapid rates of flashing, two lights are seen in place ("simultaneity"—top panel); at the proper slower rate, a single light appears to move from its first location smoothly and continuously across the screen to the second location ("optimal movement"—bottom panel); and between these two rates, a light seems to move part-way across the screen, disappear, reappear at a more distant point, and continue on to the second location ("partial movement").

of view too quickly for its contours to be made out. Beta motion, on the other hand, refers to the perception of a well-defined object moving smoothly and continuously from one location to another, analogous to the slow passage of a real object across the field of view. "Optimal motion" is synonymous with beta motion. I shall use the terms phi, beta, and optimal motion in these senses. Recently, however, Kinchla and Allan (1969) have used the term "phi" to refer to still another aspect of motion perception, one unrelated to the terms discussed here. Their usage should not be allowed to confuse the issues further.

Phi motion was given the greatest theoretical attention because its occurrence seemed to prove conclusively that motion is not a complex perception built up out of a sense of temporal order and memory of an object's location. In summarizing his results Wertheimer wrote,[2] "The sense of motion is not constructed in any material way from the subjective filling in of objects in the intervening places [between the flashes]; . . . the phi phenomenon [is] separable from the [processes mediating the] appearance of the two stimulus objects." Moreover, as to its relation to real motion, "The perception of motion from the successive stimulation [of two separate points] appears equivalent *as motion* to the perception of motion that occurs when a physically moving object is exposed. . . ." For Wertheimer, then, object and motion were distinguishably different aspects of perception, for one could have either without the other. They differed not only in appearance, but in their functional organization by the nervous system. He therefore rejects all theories that assume that motion perception depends upon fading traces in neighboring retinal regions or upon comparisons in memory of objects in different locations. In addition, he equates, functionally, real and illusory motion.

In his experiments Wertheimer measured the duration of the flashes and the interval between them that yielded good perceptions of motion. His stimulus flashes were usually quite brief, but his measurements nevertheless confounded variations in stimulus duration with variations in the interstimulus interval. The three observers for whom he presents data differed somewhat, but all of them tended to report good motion when the interstimulus interval was in the region of 60 msec. Later, Korte (1915) and more comprehensively Neuhaus (1930) manipulated these variables in a systematic way, with results that support but qualify Wertheimer's. In the 1920s and 1930s a number of American investigators fastened on the value of 60 msec as if Wertheimer had asserted it to be a natural constant, and showed that it was not. Some investigators still try to measure *the* optimal interstimulus interval, without appreciating its dependence upon many other variables.

Another theory prevalent at the time, espoused by Helmholtz but

---

[2] "Der Bewegungseindruck ist materialiter nicht in subjektiver Ergänzung von Zwischenlagen des Objekts konstituiert; . . . das φ-Phänomen in Abgelöstheit von den Erscheinungen der beiden Reizobjekte" (p. 235). "Die bei Sukzessivreizung resultirende optimale Bewegung zeigte sich in bezug auf die *Bewegung* als gleichwertig dem Sehen der Bewegung bei Exposition eines entsprechend wirklich bewegten Objekts . . ." (p. 236).

refuted by von Szily (1905) among others, was that certain impressions of motion are attributable to kinesthetic stimulation associated with movements of the eye in following an object. Wertheimer tested this idea by presenting in one flash two lines which were a small distance apart and in a second flash two others flanking the first two but a larger distance apart. The perceived motion then went in opposite directions simultaneously, hardly possible if motions of the eye were required for perceived motion of objects. Again, he presented one flash to one eye and the second flash to the other and achieved good perceptions of motion; he alleged therefore that motion perception was "behind the eye" and not "in the eye". He found that the optimal interstimulus interval varied with the spatial separation between the flashes, and that some temporal overlap of the flashes was tolerable without interfering with the motion percept. Korte (1915) took on the task of varying space and time systematically, getting results that Koffka later formalized into "Korte's Laws"; the overlap effect was studied systematically some years later by McConnell (1927). These results all clarified the role of timing of the flashes in producing the phenomenon, and the emergence of motion out of succession and simultaneity.

In other experiments Wertheimer made the color of the origin different from the color of the terminus and found that observers reported the flashes to change color in flight. This result was later confirmed by van der Waals and Roelofs (1930, 1931) and by Squires (1931).

Wertheimer also studied the effect of set and attitude, using two lines that formed an angle. If the two lines are drawn so that an oblique line intersects a base, the motion perceived when they are alternated is typically over the shorter distance, that is, through the acute rather than the obtuse angle. When he increased the angle on successive exposures, however, observers continued to report motion over the "familiar path", now an obtuse angle, even when the angle formed was as large as 160° (Fig. 2.2). Moreover, if enough antecedent trials had occurred, motion was seen for a few additional trials even when only one flash was presented; the visual system perseverated in its response to a single flash despite the absence of its partner. And finally, among many other experiments, he abutted a small segment of a line to one point on the arc of the illusory motion and found not even a momentary lengthening of the illusory line.

This is only a short list of the experiments reported, but it covers the major topics. The paper had an enormous impact on experimental psychologists both because of the phenomena discussed and because of its special, phenomenological approach to experimentation. Regretably, the paper is not available in a complete translation, and in any

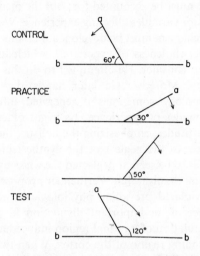

FIG. 2.2. Lines *a* and *b-b* are presented separately at a rate yielding optimal motion. The usual path of apparent motion is over the shortest distance. Gradual increases in the angle formed by *a* and *b-b*, in successive sets of exposures, enable the observer to see motion over the longer, but now familiar, path. After Wertheimer (1912).

case is sometimes difficult to read. As Shipley (1961) remarked about his own very useful extract, "This paper is particularly difficult to translate because of Wertheimer's deliberate use of words and phrases in a novel manner, i.e., as symbols of the event (e.g., 'Stationary-position-character') rather than as simple names or descriptions." In other words, Wertheimer's phenomenological approach to perception was expressed in his highly nominalized descriptions as well.

What continuing issues did Wertheimer raise? Some relate to philosophy and some to facts. Consider the philosophy first. Here the issue is,

"What is the proper approach to take to the study of perception?" Wertheimer's answer, formulated more specifically some years later (1923), is that perceptual experience is based on objects, not on sensations. Hence the proper approach is to examine the constituents of object perception, and the factors that affect their relations. The approach is phenomenological in arguing that normal perceptual experience is what must be accounted for; but the method is analytical in studying the major variables affecting experience. Wertheimer argued also that some mechanism must be developed on a theoretical level that accommodates the wholeness of experience, which introspective sensationalism did not consider. He attempted to do this himself, but the results were faulty. Moreover, in later hands the originally vivid Gestalt phenomenology sometimes degenerated into mere lists of appearances and modes of experience. As it got older Gestalt theory also rigidified its attitudes, emphasizing the holistic, the immediate, the given in experience, to the neglect of the synthesizing operations. In emphasizing what is to be seen, it neglected the whereby and how come.

As to facts, the mechanism that Wertheimer proposed to account for his observations violated prevailing physiological knowledge; it also vitiated the impact of an important distinction he had drawn. He proposed that stimulation of a retinal region induced a circle of excitation in a corresponding region of the cortex. When two such circles of excitation occurred within the proper temporal interval, a fusion or "short-circuit" occurred between them, which was the physiological analog of the perceptual experience of displacement of an object.[3] In this process timing is of course critical, and therein lay the importance subsequently attached to the interstimulus interval. When that interval is too short, the two excitations would not interact (why they would not is not clear) and stationary simultaneity of flashes would be seen. When the interval is too long, again the two circles would not interact and succession would be seen. At the proper interstimulus interval the two circles of excitation would interact and motion would be seen.

This account violated contemporary knowledge; there was no evidence for such transverse processes in the cortex. Contemporary neuro-

---

[3] These circles of excitation and their interaction bear some conceptual similarity to notions of irradiation advanced by Pavlov a few years before (Kalish, 1969), and are questioned by analogous experimental procedures.

physiologists have found, in fact, that ephaptic conduction, transverse conduction across nerve fibers, is weak at best and restricted severely in the distance over which it can occur (Grundfest, 1967). The hypothesis nevertheless was later formalized by Köhler (1923) as part of his bizarre identification of visual phenomena with late nineteenth century electrodynamics, wherein the brain was regarded as a volume conductor having, in Hebb's (1948) choice phrase, all the fine structure of a bowlful of porridge.

In addition to the bad physiology, Wertheimer's proposal blurred the distinction he had drawn so carefully between object perception and motion perception. If the short-circuit was the analog of motion, what happened to the figure? And if the short-circuit was of both figure and motion, what happened to the distinction between them? The distinction between the two aspects of perception was reintroduced by DeSilva (1929) but he used a difficult terminology involving vehicles or carriers of motion and the motion itself; his ideas were not carried forward by others. For both Köhler (1923) and Osgood (1953), who proposed variants on Wertheimer's conjecture, figure and motion seem to be part of the same perceptual process. Thus, since Wertheimer's paper, little effort has been made until recently to dissociate these two constituents of the perceptual experience.

His theory enabled Wertheimer to make another radical assertion: that the mechanism of real and apparent movement is the same. He reported that under optimal conditions they are perceptually equivalent, that is, that good apparent movement looks just like real movement. By asserting that motion perception is produced by the transverse interaction of regions of the cortex, his theory made it irrelevant that real motion stimulates adjacent retinal regions while apparent motion does not. This assertion too has generated much discussion. Curiously, he reports in detail how experience, practice, and instruction affect the observer's ability to perceive illusory motion; no equivalent degree of instruction or practice is required for the perception of real motion, nor do fixations of the eye affect the two perceptions in the same way. Yet except for "posture of attention", these variables are not accommodated by the theory. As for attention, Wertheimer argued that directing one's attention to a particular region of a display facilitates the ability of the corresponding part of the visual system to become excited, and hence to generate the necessary radiations. Implicitly, then, some extra-perceptual event, the directing of attention, was thought to affect the brain's ability to generate a perceptual event, but this confusion too is neglected in the theory.

To close on a positive note, however, one should realize the import-ance to be attached to Wertheimer's treating an illusion seriously. In his day, when naive realism was still a popular epistemological assump-tion among scientists, illusions were often thought of as annoying artifacts or failures of an embarrassed visual system. "Naive realists" believe that we perceive the world as it is, directly, and that the role of sensory systems is to deliver true copies of the world to the mind. The "subjective perversions" that Külpe complained of were the ill-formed or transitory influences of expectations, wishes, or beliefs that marred the normal functioning of apparatus in its "objective" mode.

Wertheimer challenged this notion and demonstrated its inaccuracy when he showed that illusory motion could occur in a number of unex-pected and unwilled ways, in opposite directions simultaneously, for example. In addition, he showed how the phenomenon varied with variations in stimulus conditions. Thus, by elucidating some of the variables that affect the phenomenon, thereby bringing it under the experimenter's control, he showed that the illusion of motion is as respectable a psychophysical event as is any other experimentally manipulatable perception, despite its illusoriness, and therefore that its occurrence was not a matter of "subjective perversion".

Indeed, one of the important achievements of the Gestalt psycho-logists, we now realize, was the demonstration that the study of illusions revealed not defective or ephemeral processes in the visual system, but fundamental aspects of its normal functioning. Most contemporary investigators, irrespective of their relation to Gestalt psychology, agree with this view; with the possible exception of J. J. Gibson (1968), most recognize that all perceptions are subjective representations of stimuli. The questions we wish to answer concern the means by which stimuli are transformed into perceptions, and since the Gestaltists all percep-tions have been legitimate objects of study.

In summary, then, Wertheimer's paper exemplified a new approach to visual phenomena, a serious (but incorrect) theory to account for one interesting effect, a substantial number of measurements, and some epistemologically important assertions. Among the last are the justification for studying illusions, and the consequences of the discovery that peculiarities attend the interaction of veridical and illusory perceptions. These peculiarities should have warned Wertheimer that the perceptual equivalence or look-alikeness of illusory and veridical motion cannot carry much theoretical weight, but the warning was lost in the controversy surrounding the theoretical account he proposed.

A very large number of papers was published subsequently, first in Germany and then in other countries, investigating various aspects of the motion effects Wertheimer had described. Several types of illusory motion were identified and, following Wertheimer, given Greek letter-names; so many in fact that LeGrand (1967) complained that they used up almost the whole alphabet. The complaint is warranted in some sense because many of the phenomena were little more than alternate forms of each other, created by small variations in apparatus or procedure. Of the great number named a few have remained: beta motion, object motion across the physically empty intervening space; delta motion, beta or phi motion directed toward the first flash, found when the intensity of the second flash is sufficiently greater than the first; gamma motion, the apparent expansion at onset and contraction at offset of a single flash of light; and Wertheimer's phi motion.[4] Changes of perceived shape and color were reported by Wertheimer, when the flashes were of suitably disparate objects, and a depth percept as well when another object appeared in the path of an object in illusory motion. Koffka (1931) reviewed many of the findings of the early workers; four other reviews appeared in a short span of time, three of them in the *Psychological Bulletin*, describing specific experiments (Squires, 1928; Ewert, 1930; Hovland, 1935), and the fourth (Neff, 1936) a more theoretical treatment. Aarons (1964) recently published a bibliography of many kinds of apparent movement research, and the work still continues (for example, Enright, 1970). To this day, however, no thoroughly satisfactory theoretical account of the phenomenon has been put forth.

In the absence of a satisfactory theory, it is useful to consider what the phenomenon consists of and the conditions that a theory must account for. The facts are simple. Two properly placed and properly

---

[4] Bartley (1941) believes that the mechanism for gamma motion is markedly different from the mechanism for phi and beta, and is probably right.

timed flashes induce the illusion of a single object moving from its first location across the intervening empty space to its second location, where it may either disappear or return to its first location. If the interstimulus and intercycle intervals are equal and of proper duration, the illusory object is seen oscillating in smooth motion; if the intercycle interval is several times the duration of the interstimulus interval, the object disappears at the second location and movement recommences at the first. In other words, when conditions are right the visual system creates a perceptual object in the intervening space where physically there is none. The perceptual object created, moreover, resolves differences in appearance between the two physical objects, such as differences in color or shape. Hence the perceptual construction is not a mere redundant filling in of the space between the flashes with copies of the flashes themselves; it is an active resolution of their difference. Any satisfactory theory therefore must account, first, for the perception of motion itself from discrete sequenced flashes; second, for the perception and resolution of figural disparities; and third, for the relation of illusory to veridical motion. The remainder of this monograph deals with these issues.

CHAPTER 3

SOME BASIC MEASUREMENTS

ABSTRACT
Some of the standard data on motion perception are reviewed in light of the implications of Korte's Laws and of the similarity between real and illusory motion. It is shown that an important consequence of Korte's Laws is a constancy of velocity of objects in apparent motion, but the data do not support this assertion. An experiment examines the perception of motion as a function of the spatial density of stimulation; the results reveal that the perception of motion does not improve in a regular way when the space between two flashes of light is gradually filled in with other flashes. The conclusion is drawn that real motion and apparent motion represent the operation of markedly different processes.

Wertheimer's experiments established the "paradigm" within which most subsequent research was carried out. The main variables studied were the duration and the spatial and temporal separation of the flashes. Interpretively, attention was given to the status of phi or objectless motion and to the theory of the short-circuit. Although all the research was concerned with the perception of motion, far more, curiously, has studied illusory than veridical motion. A basis for this asymmetry of interest undoubtedly is the firm assertion by many experimenters that identical mechanisms mediate the two perceptions; hence studying illusory motion, it has been thought, would elucidate the mechanisms of veridical motion. As we shall see, there are significant differences between the two.

Motion perception had long had an anomalous status. Is motion, as Zeno claimed, an interpretation imposed on separately perceived objects, or is it created through the action of receptors or detectors analogous to those for light or touch, tone or taste? The arguments became thorny and involved, for even were there detectors for motion, no one could argue that the motion perception was the result only of their action.

19

We do not perceive our sensory transducers, we do not even perceive our retinas, as Helmholtz noted. What we perceive is always an interpretation of inputs provided by the transducers, and by other sources as well. Thus, all perceptions are computed or calculated, in some sense, so that both the Zenonians and anti-Zenonians are correct. What we need to do is determine what they are each correct about.

The first issue settled was that perceptions of motion do not necessarily require that an object be perceived in two places at two times, its locations compared in memory. Exner (1875) had shown that motion could be perceived at temporal separations less than those required for correct perception of the order of flashes; hence comparison of remembered positions is not the basis of a perception of motion. Wertheimer extended the argument by showing that a perception of motion could be obtained even without a clear perception of the moving object. These are anti-Zenonian arguments, on the side of those claiming that the visual system perceives motion "directly".

The pro-Zenonian arguments were made by Titchener (1902), by Linke and by Hillebrand (Neff, 1936). Titchener, for example, wrote (pp. 178 f.): "Our idea of movement is made up, in part, of the ideas of an object in different positions . . . [and of] the persistence of sensation after the cessation of stimulus. By the help of an after-image or of memory we are able to perceive an object, as it were, in two places at once: in the place it has just left, and in the place to which it has just come." Related ideas are expressed in more recent research that emphasizes the importance of memory (Kinchla and Allan, 1969) and of figure in the perception of motion. This issue will be discussed in later chapters. Another kind of argument that might be thought of as pro-Zenonian is based on the relativity of perceptions of motion. Duncker (1929) found that its frame of reference conditioned the perception of an object's motion, as when the moon, for example, seems to weave in and out among stationary clouds. To regard his research as pro-Zenonian would be an error, however, for his research was concerned with the interpretation given to motion signals and not with whether motion is a direct or derived perception. This emphasis upon interpretation can be found in the pro-Zenonians generally, however. Thus, the anti-Zenonians are correct when they maintain that motion is responded to directly by the visual system and is not merely inferred from comparison of memories; the pro-Zenonians are wrong, as will be shown, when they maintain that one must see a figure in two locations in order to see its motion, but they are correct when they maintain that one interprets the signals from disparate locations as motion. So let us now turn to the research. This will take us into studies of some relations affecting thresholds and supra-threshold perceptions of motion, real and illusory.

Parameters of Illusory Motion

Convinced of the directness of motion perception, the early Gestalt investigators set out to establish the laws that governed its occurrence. Koffka's student Korte (1915) measured the temporal and spatial requirements for apparent motion, with results that are sometimes too generously called Korte's Laws. They have for long been taken as basic data for the perception of motion, but the "laws" as Koffka (1935) himself pointed out are of limited generality .

Korte's results have actually been expressed in several forms; even their number is uncertain. A basis for this ambiguity is the fact that Korte treated as a single variable, "stimulus intensity", anything that contributed to a figure's salience or impressiveness, such as its luminous energy, size, or figural detail. Some modern authors follow in this tradition (Bartley, 1941; Boring, 1942) and some separate the energetic from the figural component (Graham, 1951). Energy and figure detail are said to interact similarly with the spatial and temporal variables that affect the illusion; hence three rules summarize the argument. Following Boring, they are:

1. With interstimulus interval (ISI) fixed, spatial separation of the flashes (S) and their intensity (I) are directly related: $S \sim I$ and $I \sim S$.
2. With spatial separation fixed, interstimulus interval and intensity are inversely related: $I \sim 1/ISI$ and $ISI \sim 1/I$.
3. With intensity fixed, interstimulus interval and spatial separation are directly related: $ISI \sim S$ and $S \sim ISI$.

Korte made his measurements in a curious way. First he arranged conditions so that the experimental subject reported good apparent motion. Then he changed one of the three variables and measured the condition of another that was required for the perception to be restored. The three rules, formulated somewhat like Ohm's Laws of electrical circuits and the pressure–temperature–volume relations of gases, imply that apparent motion is perceived only at particular values of the three variables. The first and third rules say that the illusory object is perceived at a constant velocity; and the second rule says that the visual system requires a fixed amount of time (the interval from the onset of the first flash to the onset of the second) to process the perception of motion over a given distance.

Neuhaus (1930) was one of the first to challenge these assertions with systematic experimentation. Unlike Korte, Neuhaus used the more conventional method of limits for his study. His procedure, in general, was to establish a set of conditions and then to vary the interstimulus interval from trial to trial, recording his observer's reports after each trial. His experiments have their own weaknesses visible to the modern eye, but they are somewhat better than Korte's. He challenged, first, the notion that apparent motion is seen only at specific intersections of the three variables; and second, that an onset-to-onset rule reliably characterizes the results.

Figure 3.1 is taken from Neuhaus's table 5. It shows the interstimulus intervals at which simultaneity gave way to good motion (lower curves), and motion in turn gave way to succession (upper curves), for three durations of the flashes at different spatial separations. Increasing stimulus duration decreases the interstimulus interval at which motion is first seen and the interval at which it gives way to succession, as Korte implied. The main fact, however, is that good motion is seen over a wide range of interstimulus intervals for any duration of the flashes and any spatial separation. (The temporal range constricts however as spatial separation is increased.) According to these data, two lines of light that are presented for 45 msec each and that are 1.4° apart can be seen in good motion during interstimulus intervals ranging from about 75 to 250 msec; conversely, when the interstimulus interval is 100 msec, their distance apart can range between 0.5° and 2.3° and motion still be seen. These results are hardly consistent with the constancies Korte described. Korte however spoke of onset-to-onset relations. Hence, the data of Fig. 3.1 can be redrawn to sum the duration of the first flash and the interstimulus interval. If there were a reliable onset-

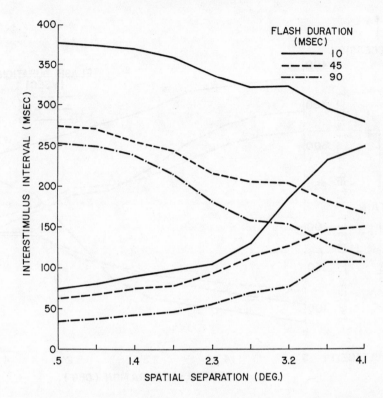

FIG. 3.1. The interstimulus intervals at which optimal motion is first seen (lower curves) and then gives way to succession (upper curves) for flashes of three different durations. From Neuhaus (1930).

to-onset interval, each set of three curves of Fig. 3.1 would superimpose when drawn in the way described. Figure 3.2 shows that the summation does not hold exactly.

Neuhaus's data can be replotted to show that velocities are not constant either. The curves of Fig. 3.2 have been redrawn in Fig. 3.3 to show the computed threshold velocity of motion as a function of the spatial separation between the flashes. The two sets of curves reveal that at any spatial separation, motion is seen at the range of velocities

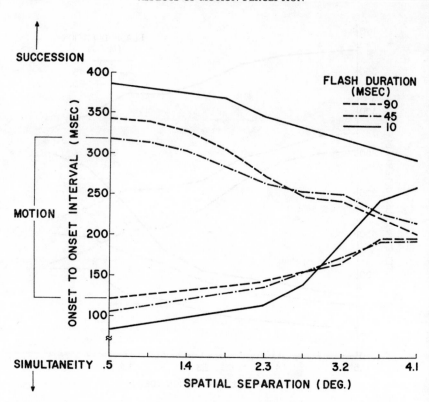

FIG. 3.2. The data of the preceding figure are redrawn to sum, on the ordinate, the duration of the first flash and the interstimulus interval separating it from the second (onset-to-onset interval).

between them; in addition, velocity increases as spatial separation is increased. There is no constancy of velocity of objects in apparent motion.

Neuhaus's subjects varied in the extent of their practice, and other conditions were also varied. His data reveal that velocity is not constant, but it is impossible to say, from his data alone, exactly how perceived velocity does vary with stimulus conditions. Even if his data were thoroughly reliable, we would still be somewhat unsure how best to

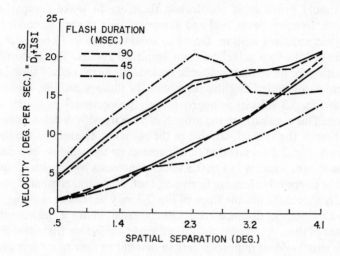

FIG. 3.3. Calculated velocity of the lights whose motion is illustrated in Fig. 1

calculate velocity. In mechanics, velocity is calculated as the distance travelled divided by the duration of travel: $V = S/T$. Some psychologists have attempted to establish equivalent rules for subjective velocity, but a serious problem hinders this effort. Spatial extent, S, can be established readily; the question is what value to use for T. Speaking strictly, T should be equal to the interstimulus interval (the travel time of the visible object) but making it so yields absurdities: apparent motion can be seen when the interstimulus interval is zero and even when the first and second flashes overlap in time (negative values of T). Both conditions yield uninterpretable values of velocity. Alternatively, one can set T equal to the interval between the onset of the two flashes; this in fact is the procedure that was followed for Fig. 3.3. But as Koffka (1935) pointed out, attributes of the second flash sometimes affect the entire perception of motion; hence we might consider setting T equal to the onset-to-onset interval plus a percentage of the second flash. We might consider many other alternatives as well, for the fact is that we do not know at present how best to calculate the perceived velocity of the illusory object.

It should prove most worthwhile therefore to make comparisons between the velocities of real and illusory objects at various conditions that yield apparent motion. Doing so would enable us to establish the relation of the two velocities from empirical results. The question is interesting for more than academic reasons. Its answer could illuminate the processes underlying the perception of illusory motion. Consider that in Fig. 3.3 velocity is approximately proportional to spatial separation. This regularity in the growth of velocity with distance suggests that time is the critical variable in the equation. Examining the lower curves in Fig. 3.2 reveals that time increases on the ordinate by a factor of about two, whereas in Fig. 3.3 velocity increases by a factor of about ten. The temporal values are fairly constant, therefore, compared to the velocity values. Hence the slope of Fig. 3.3 may be read as showing that large variations in distance are accommodated within a near-constant amount of time. As a consequence, we might conjecture that some event in the visual system requires a certain amount of time to be processed; variations in perceived velocity are merely expressions of that time distributed over the space occupied by the two flashes. Movement, not the speed of movement, is primary; speed is only a resultant characteristic of the processing.

The idea here is that the visual system requires a certain amount of time to process the flashes and that velocity is merely a reflection of the perceptual work done during that interval. One of my own experiments speaks to this point (Kolers, 1964). Two flashes of light were exposed at various durations and interstimulus intervals. After each pair of flashes the observer reported whether he had seen smooth continuous motion. Figure 3.4 shows the results for flashes whose duration varied between 24 and 215 msec. As flash duration is increased, the probability of reporting optimal motion at shorter interstimulus intervals increases, as Neuhaus also found. When these data are replotted to include on the abscissa the duration of the flashes and the interval between them, an interesting relation emerges. The separate curves in Fig. 3.5 are each for the separate flash durations that were used, and the broad curve connecting them is a hypothetical "figure-formation" function. The conjecture is that some single process is stimulated by the flashes, but its magnitude varies with flash duration. If the process is initiated by a flash of 24 msec, it starts up but then fails rapidly; if it is initiated by a longer (that is, more energetic) flash, it develops more fully. Here some sort of operating time seems to be revealed. Specific comparisons of perceived and physical velocity might lead to a clarification of the role of time in the perception of apparent motion, and especially of whether velocity, as here conjectured, is only an outcome attributed to the consequences of other processes.

The curves of Fig. 3.3 taken from Neuhaus are quite shallow. One reason for this is that space and time are not processed equivalently in the visual system. Koffka (1935) noted that in Korte's data a trebling

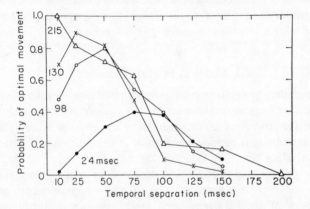

FIG. 3.4. The likelihood of seeing motion between two flashes whose durations and interstimulus intervals were varied. Duration is shown for the separate curves in msec. Adapted from Kolers (1964). Copyright © 1964 by Scientific American, Inc. All rights reserved.

of spatial separation required less than a doubling of temporal separation to maintain a perception of movement. The ratios are even larger in Neuhaus's data. Measurements on the scaling of space and of time have been made by Mashhour (1964), who found that judged spatial extent matches physical extent closely, but judged time does not match measured time closely. Judged space and judged time, that is to say, do not increment at the same rate. This fact adds a further complication to the calculation of velocity. It is likely too that the effects of stimulus energy and stimulus detail—confounded in Korte's data—also increase at different rates. What this comes down to is that space, time, energy, and figural detail are related to each other far more complexly than Korte assumed; indeed, probably nonlinearly.

Despite these severe restrictions, Korte's rules do express some vaguely accurate truths. But the specificity of the laws is grossly misleading. They were offered and have since been used to support the idea that certain perceptual constancies were present in illusory motion. These assertions are not borne out on examination. Moreover, as we shall see below, the times measured differ sharply among people, and the perceptual characteristics of the illusion seem to be resolved in too many different ways for one to believe that velocity, distance, or even figure are of primary importance to the visual system's operations, hence the basis of efforts toward maintaining constancies.

Veridical Motion: Velocity

More studies by far have been undertaken of illusory than of veridical motion. One reason surely must be the complexity of the apparatus required for studies of objects traversing large visual extents; another reason may be that motion perception is so sensitive in humans that many variables affect it. Whether the eyes are moving or stationary, whether the motion occurs on a textured or plain background, its duration of observation and the like, all affect measurements. In addition, many of the variables, such as context, are resistant to being dimensionalized, and others, such as viewing distance, have an uncertain status. Thus not very much is known about the phenomena although many phenomena are known.

The theme running through much of the research on veridical motion is derived from Zeno and is still whether motion is perceived "directly" as a primary psychological attribute of the visual system, or whether it is inferred or is a calculated or secondary attribute based on memory for position.

One argument for the primacy of motion detection has been made by Bouman and van den Brink (1953). They maintain that motion perception requires only that a certain number of quanta of light be absorbed by the retina within a specifiable temporal and spatial interval. These absorption rates are alleged to determine the motion perception. Pollock (1953), at about the same time, measured the luminance thres-

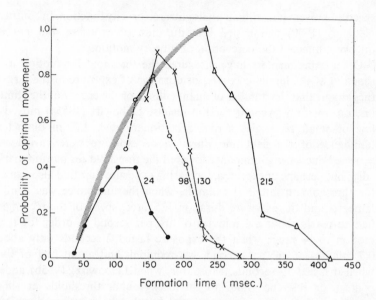

FIG. 3.5. The duration of the flashes and the interval separating them (onset-to-onset interval) are summed on the abscissa. The light curve is the hypothetical generator function for apparent motion. Adapted from Kolers (1964). Copyright © by Scientific American, Inc. All rights reserved.

hold for a small spot of light that travelled at various speeds across a screen 20° wide. The more rapidly the target moved, the greater its intensity had to be for the observer to detect it; on logarithmic paper, threshold luminance increased linearly with an increase in speed in the range from 50° per second to 1600° per second. At 50° per second the target traversed the screen in 400 msec, and at 1600° per second it traversed it in 12.5 msec. Pollock's results reveal therefore an intensity–time reciprocity, such as Bouman and van den Brink would expect, extending through the relatively long interval of 400 msec; but the results say nothing about the actual perception of motion. In fact, it is not clear that motion is even important to the data obtained; the results are probably not different from what would be obtained with stationary objects whose intensity and duration were varied. R. H. Brown (1958) also found that threshold luminance depended upon speed of movement, with an analogous reciprocity function. In contrast, Neuhaus (1930) maintained that duration of exposure, but not its intensity, affected the perception of illusory motion.

Other experimenters have measured the threshold for motion as a function of the luminance, size, or duration of exposure of the target; they report that the number of quanta alone, or the energy of the stimulus, do not fully account for the results. Leibowitz (1955a) moved a line of small rectangles across a stationary disk 3.2° in diameter; luminance of the field and duration of exposure were varied. As expected, increasing luminance lowered the threshold for motion markedly; but perhaps unexpected, increasing exposure duration also caused the threshold to fall. For example, when the luminance was 5 millilamberts and the exposure duration 125 msec, speed of the rectangles had to reach about 6.4 minutes of arc per second in order for their motion to be seen; when the exposure lasted 2 seconds only about 0.7 minutes of arc per second was required, and about half of that when it lasted 16 seconds. Moreover, when Leibowitz (1955b) added context or reference lines to the background, thresholds at short durations of exposure were unaffected, but thresholds at long durations of exposure fell markedly.

From these results Leibowitz argued that motion is processed directly (photochemically) at very short durations of exposure, but is a matter of inference or computation at longer durations. Cohen (1964), in his detailed analysis of the Aubert–Fleischl paradox, also concludes that the "physiological" and the "psychological" are two separable components in the perception of motion. This is hardly an explanation, however, because it mixes levels of interpretation. One attempts to account for motion perception at the photochemical level or one does not. Psychology and physiology refer not to different phenomena but to different ways of explaining phenomena. One cannot profitably mix the two levels. For example, Mates and Graham (1970) have recently confirmed that threshold velocity varies with the size of the moving object: trebling the length of a narrow rectangle increased its threshold velocity by an almost proportionate factor. Should this be considered a primary or a secondary, a physiological or a psychological characteristic of motion perception?

In point of fact, it may be intrinsically impossible to measure an absolute threshold for motion. The threshold one measures will always depend upon the nature of the moving object and the observer's frame of reference or task set. One aspect of the perception of veridical motion may indeed be a wave of stimulation crossing a region of the eye, but the geometry and exposure duration of the object inducing that wave play a larger role than might have been expected in the perception of motion; the response to these factors, moreover, is likely to be nonlinear. The threshold for motion does not have a fixed value. Its full description is even difficult to conceptualize.

J. F. Brown (1931), working in Köhler's laboratory, undertook such a description. He was actually the first investigator in modern times to bring the study of visual speed under systematic laboratory control. His apparatus was designed to study the perception of physical speed, but inadvertently introduced illusory motion, flicker, and other contaminating effects. The device consisted of two drums on each of which a continuous roll of paper presented black figures of various sizes and spatial separations; the aperture through which the figures were seen was also variable. The subject's task was to match the velocity of one of the drums, when the size or intensity of its figures was varied, to the fixed velocity of the other, turning his eyes back and forth between the two (which were not simultaneously visible) and calling for adjustments. Sometimes the experiments were carried out in a dark room with the sheets of paper illuminated by dim bulbs, and sometimes in a lighted room. The results differed in the two conditions. (Aubert had reported in 1886 that allowing a subject to see the stationary supports of a rotating drum lowered his motion threshold by a factor of about 10.)

Brown's theory was isomorphism run wild. He proposed that perceived velocity, like physical velocity, could be characterized by the relation $v = s/t$, the lower case letters indicating the perceptual aspect, perceived velocity, perceived distance, and perceived time. Thus he proposed $V = kv$, or that physical velocity is a measure of perceived velocity modified by a constant appropriate to viewing conditions.

Brown's experiments are complicated and neither the results nor their optimal representation is always clear (Köhler, Wallach, and Cartwright, 1942). There is some suggestion indeed that the velocity matches that subjects made were based not on velocity itself, but on the flicker and rate of disappearance created by the moving shapes (Smith and Sherlock, 1957), although this is not certain (Cohen, 1964). More-over, thresholds, as noted above, vary with the duration for which the subject is allowed to look (Leibowitz, 1955a; Goldstein, 1957); and when looking time is kept constant, with the distance over which the movement takes place (Mandriota, Mintz, and Notterman, 1962). They vary also with whether the eye follows the moving object ("pursuit movements") or remains fixated (Aubert, 1886), with the size of the object (Mates and Graham, 1970), and with the presence of reference marks (Leibowitz, 1955b).

Except for motions of the eyes, the variables refer to space and time which, of course, are components in the velocity equation. It does not matter in mechanics whether an object in uniform motion moves through 10 feet, 10 yards, or 10 miles; the calculation of velocity will depend upon the distance divided by the time, and the curve that plots velocity against time or distance will have a constant slope. In percep-tion, however, calculated threshold velocity varies in non-uniform ways when time or distance is varied, as will be shown, and even when luminance is varied (R. H. Brown, 1958; 1961). J. F. Brown's thesis clearly is too simple minded.

Mashhour (1964) tried to resolve some of these difficulties by analyzing judgments of the constituents of velocity perception, the space and time that go into the computation. He exposed a small object for various durations as it moved at various speeds over various distances. In each case the observers scaled the distance or the time numerically. The scaled values were then plotted against the physical values to show how estimates of distance varied with physical distance and estimates of time with measured time. The point of this exercise is that perceived velocity can be characterized by the expression

$$v = k \frac{s^p}{t^q},$$

and only if $p = q$ can perceived velocity be related directly to calculated velocity. Mashhour found, however, that the exponents for calculated velocity in the scaling experiment varied with conditions from 0.63 to 0.94, and were almost perfectly correlated with and almost exactly equal to the exponents for estimated time. Even in this scaling experiment, therefore, duration of observation was found to affect the judgment, rendering altogether impossible the simple equation between physical and perceived velocity.

Clarity

Still another aspect of motion perception concerns the clarity of perception of the moving object. DeSilva (1929) reported that when lights moved at speeds greater than about 50° per second, observers no longer saw a moving object; rather they reported seeing the entire 3° viewing area filled with "a sheet of light". Pollock (1953) found analogous effects; when the target moved at speeds greater than 50° per second, observers could not even tell the direction of its movement (vertical or horizontal, leftward or rightward). Instead, subjects said "that their judgments of direction of movement were based on the time difference in the appearance and disappearance of the target, not on the actual perception of target movement" (p. 453). Thus while one can *consider* perceived velocity as the equal of distance divided by time, subjects seem to use the latter two variables themselves in their judgments rather than judging velocity directly.

Smith and Gulick (1956, 1957) undertook a more direct test of the limitations on visibility of the moving object as a function of its speed. A black rectangle moved through a slot either smoothly or after first pausing at the entrance or later at the exit for a controlled interval of time. Observers reported only whether the rectangle appeared sharp and unblurred. Sharpness was reported only when the speeds were less than 15° per second. However, if the rectangle was allowed to pause at the entrance, sharpness of contour during subsequent movement was maintained through speeds that produced blurring in the absence of the pause. The curves in Fig. 3.6 show the durations for which the shapes had to be presented at the entrance and exit of the slot before observers could report unblurred contours of the moving object. For example, sharp contours could be seen with no pause at speeds of 15° per second. But at 25° per second the stationary rectangle had to appear for 300 msec at the origin and at the terminus of the movement if its contours were to appear sharp and unblurred. In control experiments it was found that pausing at the beginning of movement was more important for maintaining a perception of contour than pausing at the end. If the object motion is slow enough, the visual machinery can work in "real time" to form clear perceptions of a moving contour, but at more rapid speeds the contour-forming processes are interfered with. Hence, if the perceptual machinery first gets a good look at an object, it can sustain a clear perception of the object in conditions in which blurring would ordinarily occur. The ability to see good motion even with blurred contour, however, demonstrates, as Wertheimer had demonstrated with phi

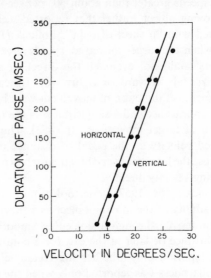

FIG. 3.6. A moving shape that appears blurred can be seen sharply and well-defined if it is first exposed in a stationary position. The ordinate shows the exposure required for shapes moving at speeds indicated on the abscissa. From Smith and Gulick (1957).

motion, that the visual system distinguishes between figure perception and motion perception, and that it requires more information to form and maintain contours than to report movements. Smooth and complete contours are seen in motion only at speeds less than 50° per second; at low to moderate luminances, such as Smith and Gulick used, the limit is about 15° per second. Object motion itself can be seen at much higher speeds, the precise value depending upon where the eye is looking and on how much supportive detail there is in the visual field.

Comparison of Real and Illusory Motion

Differences in perceptible velocity may distinguish between real and apparent motion. Real motions of sharp contour can be detected at speeds between a few minutes of arc per second and 25–30° per second, depending upon illumination and duration of observation. It is not

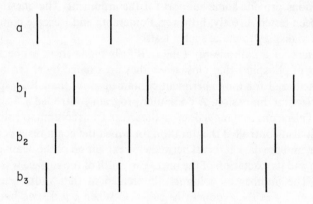

FIG. 3.7. Alternating the set of lines in *a* with any of the sets in *b* at a fixed interstimulus interval generates perceived movements of various speeds.

known whether the range for apparent motion is the same. The sets of lines in Fig. 3.7 provide a means for making some tests. Let the lines of *a* in Fig. 3.7 be the first flash and any of the sets in the remaining rows be the second. A very great range of speeds can be seen at a single interstimulus interval when the different sets are displayed. Slow speeds are seen when the distances separating the lines are small, and faster speeds when they are larger. A question that has still not been resolved concerns the lower limiting value of speed of illusory motion. Bartley (1941) conjectured that illusory motions would have a narrower range of speeds than real motions, that is, could not be made to go as slowly as real motions can. The conjecture is worth testing. I have unsuccessfully attempted to do so; the lack of success illustrates some of the difficulties involved in all measurements of motion perception.

The image of an object in real motion stimulates contiguous regions of the retina, but the image of an object in illusory motion does not. A persistent question asks whether the intermittency of stimulation associated with spatial separation makes any difference to the formation of the perceptual experience. Many theories of apparent motion regard the spatial intermittency as irrelevant; the discovery of "receptive fields" (Jacobs, 1969) in the eye and of the eye's extensive lateral interconnections provide some support for the argument. The question has never been tested directly, however. Pomerantz and I used a computer-generated display to make such a test.

The face of a cathode-ray tube (CRT) is made from a fine matrix of spots of phosphor that glow when they are excited by electrons. The CRT face used for the experiment contained more than 1024 spots in the horizontal dimension. A computer program controlled a 5-cm high beam of electrons as it moved across the screen from column to column of spots; it also controlled the duration for which the beam rested on each column, and the dark interval between the extinction of the beam on one column and its excitation of the next. The width of the surface was about 13 cm. The number of columns illuminated in that width varied as powers of 2. Let the exponent be called k. When k is 1, two lines separated by 13 cm are illuminated, and when k is ten, 1024 contiguous lines are illuminated. The problem was to find the durations of lines, the interline temporal interval, and the spacing that yielded perceptions of smooth, continuous motion. Viewing distance was 1 meter (see Fig. 3.8).

When k is 10, the situation is analogous to real motion, for contiguous columns of phosphor are illuminated. When k is 1, the situation is the limiting case of apparent motion. As k is increased, increasingly more retinal elements are stimulated between the origin and the terminus of movement. The literature on motion perception permits at least three predictions to be made.

One prediction is that a constant velocity of motion should be obtained by increasing k and simultaneously decreasing the duration of the lines or the interline temporal interval. A second prediction is that smooth continuous motion should be seen at decreasing speeds as k is increased, duration and interline temporal interval varied accordingly. A third prediction is that Korte's rule for the interaction of spatial separation and interstimulus interval (interline temporal interval, in this case) should hold for two lines presented at varied distances. None of these predictions was borne out, however.

When k was 1, two lines appeared on the screen 13 cm apart; with proper timing, good illusory motion was seen. When k was 10, all 1024 lines appeared in succession; continuous motion across the whole screen was seen. When k was 5, thirty-two lines appeared on the screen, 4 mm apart (about 14 minutes of visual angle); smooth continuous

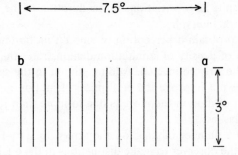

FIG. 3.8. When only lines a and b are presented on the face of a cathode-ray tube, they can be seen in apparent motion. Adding lines successively in the space between does not yield a monotonically improving perception of motion. The figure illustrates the condition for $k=4$, 16 lines on the screen, exposed in sequence from a to b.

motion was seen but its quality varied greatly. Vivid, compelling, and pleasing motions were most prevalent when speeds were in the region of $1°$–$2°$ per second (cf. Mashhour, 1969); the motions were judged less vivid and compelling at faster speeds even though the value of k remained unchanged.

However, when k was 2, 3, or 4, so that four, eight, or sixteen lines appeared on the screen, the perception of smooth continuous motion was never attained. Two lines on the screen yielded good apparent motion, and thirty-two lines did also; but the intermediate values of four, eight, and sixteen lines produced degenerate displays in which a sense of motion was always available, but not a sense of continuity. Moreover, the brightness and compellingness of the "moving" line varied markedly: the illusory line between the flashes was always paler than the flashes themselves. This sharp and rapid waxing and waning of brightness destroyed for the observers tested any acceptable equivalence between the sense of motion from two lines and from four, eight, or sixteen lines. It was never the case, in this range, that adding lines improved the perception of motion. Thus if "quality of motion", that is, smooth continuity, were rated as a function of number of lines presented, the result would be a U-shaped curve in which "quality" was rated high when two lines or more than sixty-four lines were presented, and was rated lower at intervening densities of spatial stimulation. These results held despite large variations in the duration of the lines and the interline temporal interval.

The failure to obtain a perception of smooth motion at these intermediate values of density of stimulation demonstrates that the motion system is not linearly connected. Adding stimuli in the space between the flashes does not improve the perception of motion monotonically. The perception of motion therefore is due to more than just the order and timing of flashes.

Increasing the number of lines on the screen to sixty-four or more still permitted the observer to make distinctions in the quality of movement. Often, discontinuous but compelling motion was seen as the line flashed on and off in one location, then appeared after a sufficient temporal delay in another location. The successive appearances produced a clear sense of something moving, but not of a continuously visible moving object. This sort of perception seems analogous to the conditions Leibowitz (1955) and Cohen (1964) call "inference of motion". The suggestion here, however, is that object perception and motion perception are analytically separable components of the whole, whose threshold conditions differ. What is described as inference of motion is actually a near-threshold condition in which motion signals are not quite strong enough to induce a well-formed perception, but are not entirely absent; the motion signals are "subliminal", so to speak. They may be brought above threshold by strengthening the signals themselves, and by the influence of knowledge, set, expectation, and the like, as we will see in Chapter 10. Describing the data in this way avoids the uncomfortable strategy of using two levels of analysis, the "psychological" and the "physiological", to explain a single set of observations.

In review, I have shown that certain implications of Korte's Laws and related measurements of the parameters governing apparent motion are not borne out; chiefly in dispute is the implication of a constancy of velocity of the illusory object. Rather, the evidence suggests that what is measured as a velocity is actually the attribution of figure displacement to certain perceptual operations that require time to run off.

In addition, we have seen that measurements even of real motion are often predicated on a faulty analogy to mechanics. The data raise the question whether the visual system ever notes velocity directly, and suggest, to the contrary, that some of the constituents of velocity such as time, distance, and intermittency are more important to judgment than is velocity itself. And finally, with direct attempts to measure the relation of spatial density of stimulation to the perception of motion, we made the discovery that incrementing the spatial density of stimulation does not similarly improve the perception of motion. It seems therefore that there is no necessary continuity of processing between spatially separated and spatially contiguous flashes; the ways the visual system constructs the two perceptions of motion seem to be quite different. The variability of motion judgments raises the question, moreover, whether it is useful to try to analyze and measure the perception of motion itself; it might be more profitable to study the effects of motion on other aspects of perception. Some experiments along this line will be discussed in Chapter 9. Having noted these disparities, let us turn now to studies of apparent motion itself and to the question of the relation of object perception to motion perception.

CHAPTER 4

MOTION AND FORM

ABSTRACT

One argument made frequently is that the perception of a figure is prerequisite to its being seen in illusory motion; hence, contour or shape (and especially the contour or shape of the object in the second flash) is thought to determine the characteristics of the perception. Experiments reveal that contour or shape is a relatively late stage of visual processing, and plays a dependent rather than an independent role in illusory motion; the visual system, it is shown, is responding to excitations primarily, not to figures. The Gestalt argument of the primacy of contour in perception is rejected.

Some students of apparent motion such as Linke, Hillebrand, and Stern emphasized the figural aspect of the objects whose intermittency yielded the perception (Neff, 1936). After Wertheimer, however, experimenters concentrated upon the location of the flashes and their time relations, and paid less attention to the shapes and their interpretation. Wertheimer and the Gestalt investigators seem to have been of two minds about the matter. The short-circuit theory makes no mention of figures or shapes; it talks only about radiating patterns of excitation, in Wertheimer's (1912) version, and columns of electrical excitation, in Köhler's (1923). But when Ternus (1926), under Wertheimer's direction, extended the discovery of Gestalt organizing principles (Wertheimer, 1923) to apparent motion, and when von Schiller (1933) discussed the tendency of the spatially and temporally disparate shapes to become assimilated into a Gestalt unity, figure and its organization were clearly important principles in mind. The result is that two models can be formulated from the earlier work. One model states that only a disparity between locations of stimuli is perceived and motion is created to

40

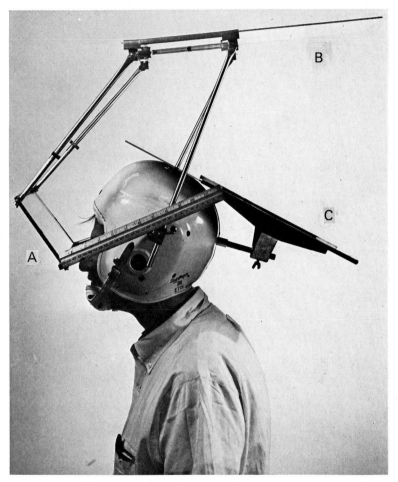

FIG. 4.1. A partial mirror in front of the eyes (A) and full mirrors above (B) and behind the head (C) produce simultaneous retinal images of the straight-ahead and straight-behind. The observer typically does not combine contours, but allows one or the other visual message to dominate perception. From Kolers (1969).

resolve it; a second model states that a disparity between figures in different locations is perceived, and motion is created to resolve that. I shall show that neither model is wholly correct, but that the weight of evidence lies with the first.

Figural Identity

The problem of figural identity that Wertheimer perceived and Ternus worked on, underlies some of the ambiguity in the Gestalt position. In motion pictures, for example, an actor's arm may sweep across his body. At any instant of time the conditions of physical stimulation make it possible for the arm to be paired in motion with many of the other contours that form parts of the body or even parts of other objects in the scene. For most observers, however, the arm and torso retain their separate identities, and a whole arm is seen to sweep across a torso without interacting with the many contours in its neighborhood. The problem of figural identity concerns just this aspect of the perceptual situation, in which despite the opportunity for local interactions, figures retain their wholeness and their identity.

The problem does not need the motion pictures for its exemplification. The maintenance of figural identity in a continuously changing stimulus flux is one of the major achievements of the visual apparatus. Why do shapes and patterns not blend and melt into each other, and what rules govern which things will be seen as belonging together and which separate? The eyes move continuously within a head and body that are also usually in movement, and the physical signals impinging on the receptors also change continuously, yet figures retain their shapes and objects retain their position in perceptual space. The problem may be illustrated with the simple device shown in Fig. 4.1 (Kolers, 1969). The mirror in front of the eyes is transparent, but the mirrors above and behind the head are fully reflecting. In a lighted environment, stimulation both from behind and before the observer impinges on his eyes simultaneously. The observer has little trouble in resolving the optical melange into its proper parts, however. Exactly the same results may be achieved with two pictures that are optically combined in a viewing box. When they are first shown, their images superimposed, the observer can make little sense of them. If one of them is jiggled about for a moment or so, however, the optical melange decomposes into two moderately distinguishable parts whose separate characteristics can then be assessed. Related observations regarding the acquisition and maintenance of figural identity lay at the heart of Ternus's research, but his experiments themselves, regrettably, do not speak to their explanation.

Ternus used a large number of displays made up mostly of points of light, some of which were exposed once and some twice in a sequence. For example, four lights A, B, C, D are arranged in a row. Three of them, A, B, C, are exposed first, followed by another three, B, C, D. The question is, will the re-illuminated points retain their locations in space, their identities, in Ternus's terms, or will they lose their local identity and become part of a Gestalt? In the case illustrated, the points lost their identity, for the observers saw what seemed to be a group of three lights move laterally as a whole unit. An object—a cluster of three lights—retained its identity at the cost of the identity of the local elements. Thus, one might infer, the figural whole dominated the local excitations.

The importance of figural whole is illustrated with another sequence. This time let six lights be exposed on each of two presentations. If they are arranged in a row so that on one trial A, B, C, D, E, F are followed by D, E, F, G, H, I, then as in the preceding example a whole cluster (a line of lights) is seen to move, the individual points again subordinating their identity to the whole. But if on another trial, A, B, C and G, H, I are in one row while D, E, F are positioned between and above them, the exposure conditions remaining the same, then D, E, F will retain its own identity on re-illumination and motion will be seen only between the successively exposed triplets in the lower row.

Ternus went on in many other experiments to show how Gestalt formation rules governing the grouping of elements controlled whether points retained or lost their identity, and how a figural Gestalt or whole took precedence over local excitations. Location of points, in his analysis, was subordinated to composition of the points in a figure, and figural factors as expressed in Gestalten dominated the perception. The implication of these findings is that the visual system assesses the Gestalt, the figure, before it decides how to construct its motion.

A related, but more clever line of analysis was undertaken by von

Schiller (1933), who also exposed points in various configurations, and whose conclusions emphasized the assimilation into a single figural unity of the successively exposed parts. Here too figural wholes and Gestalten are implicated as controlling events in the motion experience.[1]

The importance of Gestalt identity (figure) and of semantic identity (interpretation) is suggested in many other experiments: for example, Steinig (1929) and DeSilva (1926) found that "meaningful" figures are seen more easily in motion than "random" shapes. In two exposures a dog is seen first at the foot of a chair and then perched on its back (Steinig), or a hand is shown first in an upper and then in a lower location (DeSilva). Motion is reported to be more compelling and to extend over a greater range of interstimulus intervals with these presentations than with geometrically equivalent but less "meaningful" ones. Even more recent experimenters have emphasized the importance of figure in resolving the motion experience (Jones and Bruner, 1954; Jeeves and Bruner, 1956; Toch and Ittelson, 1956). At another level of analysis is Oldfield's (1948) discussion pointing out that wide deviations from the laboratory conditions required for apparent motion are found in motion pictures without seriously impairing the motion perception. The implication in all of these studies is that the visual system assesses the figure before creating its motion, as in the second model described above.

Despite this strong involvement of figure in the perception, relatively few experiments manipulated figure systematically as an independent variable. Many of the early experimenters, beginning with Wertheimer (1912) and including Neuhaus (1930), van der Waals and Roelofs (1930), and Higginson (1926), presented disparate shapes in the two flashes and found that the visual system resolved the disparity; it resolved the disparity even when the flashes were of different color (Squires, 1931). Van der Waals and Roelofs alternated a circle and a cross in place without destroying the motion experience, finding the one figure grow into the other; Neuhaus alternated an upright and inverted vee and found that the disparity was resolved in depth; and Squires confirmed Wertheimer's finding that differences in color were resolved

---

[1] I discovered von Schiller's unduly neglected paper only after Pomerantz and I had finished the work described below, to find that in many cases we had only repeated his experiments. However, our conclusions are quite different.

smoothly by the visual system.[2] The smooth resolution of the disparities seems to have fooled some of the investigators; in fact van der Waals and Roelofs carried its implication to its limits when they argued that the motion perception could not be formed in the nervous system until after the second flash had appeared, for otherwise the system would not know how to resolve the disparity between the flashes. (It was with this in mind, perhaps, that Koffka (1935) suggested that some portion of the duration of the second flash should properly be included in the computation of velocity.) As I shall show later, this seemingly plausible concern with the appearance and identification of the second flash need not be interpreted as it has been.

Shape and Motion

The suggestions put forward by van der Waals and Roelofs are perhaps the most radical expression of Model 2, emphasizing the importance of figure perception for motion perception. They did not phrase the issue in quite that way, of course; in fact, despite Wertheimer's early demonstration of the difference between figure perception and motion perception as expressed in the figureless phi-motion, relatively little effort was made to maintain the distinction.

The late Adelbert Ames (1955) used evocative language to describe "Thatness" and "Thereness" as two different aspects of perception. Of these two, the threshold for location (hence, motion), in most vertebrates, is far lower than the threshold for objects, shapes, or forms. We can detect a motion or jitter at the edge of our visual fields long before we can identify the object that is moving, and for many animals an object may be altogether undetectable until it moves (Walls, 1963). Thorson et al. (1969) have shown that movement between two closely spaced dots that are presented sequentially can be detected even when the spacing between them is too small to be resolved when they are stationary. In fact, the phi experience, objectless motion, can be described as the result of a signal's exceeding the threshold for motion but

---

[2] Regretably, no one has studied this color effect systematically. Nelson Goodman has noted the importance such a study could have in finding, for example, whether disparate colors go through gray or white, a neutral point, when they are seen in motion, and whether the formal metric of color space is preserved in the transitions.

not the threshold for shape. These differences in the visual system's characteristics were discovered and described very early in the study of illusory motion, and then were dropped. In the case of van der Waals and Roelofs, the issue was turned around, and figure perception was made almost an antecedent of motion perception, the latter alleged to be unable to occur before the former had occurred. The role of figure was not studied in anything like a systematic way until Orlansky's (1940) experiments, however.

The main purpose of Orlansky's study was to find the way that differences in the shapes of the two flashes affected the temporal characteristics of the perception of illusory motion. His measure was the range of interstimulus intervals from the emergence of motion out of simultaneity to its disappearance into succession. He measured this range when the shapes presented were identical pairs (circles, squares, and the like) and when the two members were disparate; and when the disparateness had some aspect of similarity (both shapes, while disparate, contained curves). His general findings were that the lower and upper thresholds, the range, the ease of seeing motion and its compellingness were all affected by variations in the pairs. Disparity of the shapes interfered with the formation of the motion illusion. Thus, again, presumably the visual system had to assess the disparity in order to know how to resolve it, and when the requirements exceeded the visual system's figure-resolving capabilities, the motion illusion itself was impaired.

Orlansky does not prove his case, however, for his method actually obscured the major variables of interest. He deliberately confounded stimulus duration and interstimulus interval, for he assumed that only the blank interstimulus interval played an important role in the perception. As long as the shapes were visible, he thought, their durations were not important. A confusion about the specific role of duration and interstimulus interval existed for rather a long time in the literature (DeSilva, 1928), and of course the emphasis upon the interstimulus interval lay at the heart of the Gestaltist's faulty interpretation of their own theory. Neuhaus (1930) had shown several years before Orlansky wrote that duration played an important role in the phenomenon, and that the interstimulus interval did not have a fixed value; but perhaps Wertheimer, who was one of Orlansky's advisers, suggested to him that only onset-to-onset time was important. In any case his apparatus varied only the time from onset of the first flash to onset of the second; stimulus duration was made a fixed fraction of that interval. However, if duration or interstimulus interval interacts selectively with shape, as the theory implies, then the range is useless as a measure, for it confounds the influence of the very variables whose action is the point of the study. Despite this confusion, Orlansky found that motions sometimes failed to occur when the shapes were made very different.

The finding creates its own problem of interpretation, for what do similar and different mean? What is a circle more or less of than a square and how should the difference between an arrow and an arc most properly be specified? We have no answer at the present time because we lack a generally useful notation for visual shapes, even a generally useful metric. Goodman (1968) in fact has speculated that it may be inherently impossible to devise a notational system for pictures that captures them with anything like the completeness with which we can capture music in a score or spoken language in written. Be that as it may, no generally useful notational system now exists for visual figures. Orlansky's experiments on the other hand suggest that the visual system might be used to help devise both a notation and a metric. We must take a short detour in order to appreciate the problem.

Aspects of Shapes

A persistent problem in studies of visual perception has been the absence of a convenient notational system for shapes. We can specify what we do to a shape, such as varying its intensity, location, duration, and the like; and we can name a shape in question such as "square" or "circle". But many shapes have only grossly generic names ("irregular polygon") and their specification or notation requires, say, locating them on a cartesian graph or generating long lists of binary digits that indicate excitations on a cathode-ray tube. A great deal of work directed at these problems of notation and metric has been undertaken in the recent past. Geometric features of the shapes are first measured and measurements correlated with judgments of "similarity" or with thresholds of detection. This metricizing of shapes has not proved to be an especially rewarding attack on the problem, however (Kolers,

1970); indeed, it is not even clear that judged "similarity" is sensitive enough as a measure for the purpose. The reason is that "similarity" is not a single property that can be dimensionalized directly as loudness, or sourness, or squareness can. It is, rather, a composite, the importance of whose constituents varies with the task set the subject. For example, when people are asked to assess the similarity of shapes, orientation of the shapes is a critically important variable in some judgments and not in others (Kolers and Zink, 1962). Furthermore, the notion of similarity has to be assessed in conjunction with a level of analysis. Shapes may be dissimilar at the graphemic level, as patterns, and yet be quite similar at the semantic level, as objects (Fig. 4.2); conversely, shapes can be similar at the graphemic level and be different at the semantic level, as the words *lead* and *lead* illustrate. One might turn the problem around, however, by following up Orlansky's work. Many pairs of shapes could be presented to the visual system as flashes to be seen in apparent motion, and the visual system's ease in creating motion be used to infer a proper sorting of the displays in respect to their "similarity". Pomerantz and I undertook such a study.[3]

Two kinds of pairs were presented, identical pairs and disparate pairs. The identical pairs were circles, squares, triangles, and arrows, equated for area; the disparate pairs were combinations of one item from each of two identical pairs, as a circle and square, triangle and arrow, and so on. The two shapes of a pair were presented at a fixed duration but at many different interstimulus intervals; the subjects indicated after each presentation of a pair of shapes whether they had seen smooth continuous motion between them. If figural processes were important to the perception of the illusion, we would expect that many fewer judgments of smooth, continuous movement would be made with the disparate pairs than with the same pairs. To the contrary, it was found that differences between the items of the pairs made very little difference to the perception. The observers saw one shape change smoothly and continuously into the other. Figure 4.3 shows the results obtained from four experienced observers when the two members, A

---

[3] Posner and Mitchell (1967) and Sternberg (1967) have individually suggested tha speed of reaction in judging shapes as similar or different could also be used in this way. The results reported below imply that little success would be found with such methods.

Fig. 4.2. Aspects of "similarity": the shapes are graphemically different but semantically similar.

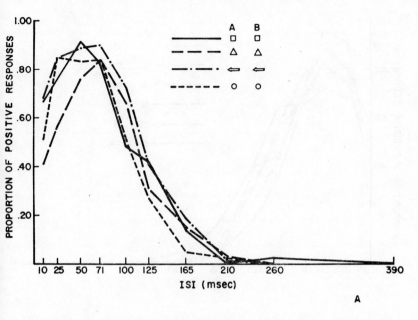

FIG. 4.3. The likelihood of seeing optimal motion between pairs of identical shapes as a function of the interstimulus interval (ISI). This and the following two figures are from Kolers and Pomerantz (1971).

and B, were identical; Fig. 4.4 shows the results from the same observers when the members of the pairs were disparate in shape; and Fig. 4.5 shows the combined results for all of the same pairs, all of the disparate pairs illustrated in Fig. 4.3, and some additional disparate pairs, called "complement", in which the arrow was the first shape and circle, square, or triangle the second.

A hasty look at Fig. 4.3 and 4.4 might lead one to infer that differences between shapes do make a significant difference to the results. Pairs of triangles seem to be seen in motion somewhat less frequently than pairs of squares, for example, and a circle and triangle less frequently than a triangle and arrow. In fact, however, as Kolers and Pomerantz (1971) pointed out, the differences varied from observer to observer, and the rank ordering of the pairs in respect to ease of seeing

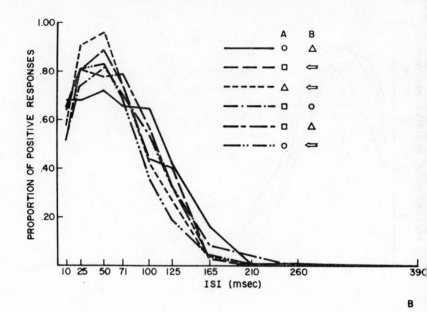

FIG. 4.4. The likelihood of seeing optimal motion between dissimilar shapes.

motion differed from testing session to testing session. When the results were treated statistically, the differences between shapes accounted for only one to three per cent of the total statistical effect (chi-square test). Thus while disparity of shapes is not altogether without effect, its influence upon the ease of seeing motion is quite small in these tests. The visual system seems to be able to move different pairs of identical shapes equally well; and it seems to be able to change disparate shapes into each other in movement about as easily as it can move identical shapes. The statistical difference due to disparity is so small as to suggest that the Gestalt emphasis upon shape as primary is too restrictive; indeed, the experiments that follow show that the emphasis is wrongly placed.

Let us pause for a moment first to describe the testing situation and the way experiments will be described. The stimuli were presented

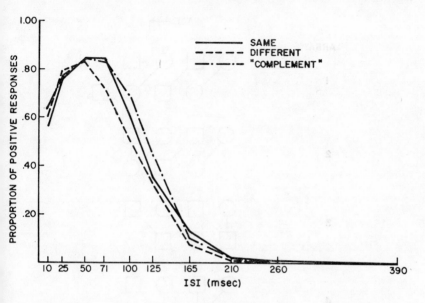

FIG. 4.5. Summary of the data for same pairs, disparate pairs, and some others.

by means of a four-channel tachistoscope that timed the presentation of two flashes and of the blank intervals between them. The presentations were usually 200 msec each, the interstimulus interval 10–15 msec, and the intercycle interval either 3.2 sec or else equal to the interstimulus interval. The visual field in which the displays appeared was, therefore, always illuminated, and the shapes displayed were black. The field measured about 7° wide and 6° high; its luminance was about 4 milli-lamberts. Viewing distance was 1 meter. In all of the tests to be reported, only well-practised subjects were used, and usually only two or three of them, hence the demonstrations should be read for their qualitative rather than their quantitative significance.

The experiments were carried out with displays composed of simple geometric figures, most of them small. A circle 6 mm in diameter (about 23 minutes of visual angle) was taken as the standard figure; it could appear in any of seven positions in a horizontal row, whose center to

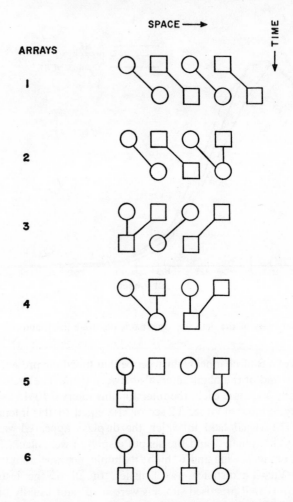

FIG. 4.6. Arrays 1–6. The shapes in the second row of each array appeared, after a slight pause, in the locations of the shapes immediately above them. The connecting lines show what movements are perceived.

center distance was 12 mm. Therefore we could have a row of seven circles, each one 6 mm in diameter, and each separated from the next by 6 mm. I will refer to the center-to-center separation of 12 mm as one figure-unit.

For apparent motion, two flashes must be presented. The two will be illustrated as two rows of figures. In these experiments the second flash was superimposed on the first. Array 1 of Fig. 4.6 shows that a row of circles and squares, each figure 6 mm wide, was followed by a second row of circles and squares, the second row displaced one figure-unit to the right; hence, every shape in the second row appeared in the spatial location of the shape immediately above it in the first row. In this notation, therefore, columns indicate spatial location, and rows indicate sequence. The thin connecting lines, in turn, show the origin and terminus of motion. Array 1 should be read to indicate that each shape seems to move rightwards; in fact, the line of shapes appears to move as a whole. If the interstimulus and intercycle intervals are made equal, a single row of shapes is seen to oscillate, in perfect agreement with Ternus's results: like figures move together.

A feature of Array 1 is that three shapes in the second row are superimposed upon three in the first row, but the leftmost circle in the first display is followed by a blank space, and the rightmost square in the second display is preceded by a blank space. Arrays 2–6 studied some aspects of that difference.

In Arrays 2 and 3 the displays continue to move as a line, but in Array 2 the rightmost square changes into a circle, and in Array 3 the leftmost circle changes into a square. Thus motion of the line of figures can occur when only one of the ends is restimulated. Array 4 shows that lateral motion is seen even when both end points are stimulated only once but the center is restimulated, although in this case the lateral motions are in opposite directions. However, Array 5 shows that the converse is not true: when both ends are restimulated, the extreme figures change shape while the interior figures merely blink on and off. Array 6 demonstrates a complete failure of lateral motion: each figure changes shape smoothly to accommodate the figure that replaces it and motion of the line as a whole is lost.

The main conclusion from these simple demonstrations is that figural identity is a weak component at best in the illusion of motion; it is

A.M P.—C

certainly subordinate to the interaction of local patterns of excitation, and to the differential influence of means and extremes. In Array 5, for example, the visual system changes the appearance of shapes at the extremes and allows the inner shapes to blink on and off; but if shape itself were important, the middle shapes of the first display should move smoothly to the extremes while the extremes disappeared. One might think that having built a square the visual system would find it easy to move that square to a new location. Instead we find that given the choice of retaining a figure but moving it to a new location, or retaining a location and building a new figure within it, the visual system performs the latter task. Figure-identity and figure-maintenance do not seem to be important activities in the visual system's construction of apparent motion in these tests. (They do exert some influence, however, as will be shown below.)

We may note, parenthetically, that in an earlier day experimenters worried a great deal about the nervous system's ability to form clear perceptual representations of objects when their images fell upon the same part of the retina within certain temporal bounds. The term "clearing-up time" was coined to express the matter (for example, Dodge, 1907), and the term has persisted in use to the present time. In the experiments just described, one may see that the notion of clearing-up time contributes nothing to an understanding of the results, for "preconscious" patterns of excitation rather than conscious perceptions of pattern dominate the results.

The unimportance of contour in these results stands in marked contrast to the assertions of the Gestalt writers and those influenced by them. The Gestaltists wrote often of givens, immediate constituents, directness, and similar concepts that treated visual experience as a pictorial replica of the retinal image.[4] Among the givens for visual perception, contour was pre-eminent, and under Wertheimer's direction Gottschaldt (1926) showed how powerful an influence contour can

---

[4] Speaking strictly, the isomorphism was postulated as between perceptual and cortical events (Köhler, 1947). The retina maps fairly faithfully onto the cortex, however (Polyak, 1957), so that an isomorphism between perception and cortical event must extend as well to retinal events. Hence, after all the hard struggle, Köhler's theory reduces to an alleged correspondence between retinal image and perceptual experience, the self-same naïve realism that Wertheimer had undertaken to demolish.

exert in human form perception. Werner (1935), although attacking the Gestalt emphasis upon givens with his brilliant series of investigations on the formation or development (rather than givenness) of contour, nevertheless made of contour the fundamental datum, the thing without which there is no perception. Even for him contour was basic. Experiments on apparent motion challenge these assertions however. They demonstrate that contour is a relatively late aspect in the construction of visual perceptions, that its formation is remarkably plastic, and that it plays a dependent rather than an independent role in figural processes. Here, interactions between local regions of the visual system seem to be more important than the figures displayed.

The visual system, having built a simple figure, does not retain it and merely move it about in space; rather, the system preserves a location and builds or modifies its contents anew. The success of the construction depends in a complex way on timing. As Kolers and Pomerantz (1971) pointed out, increasing the duration of the first flash actually enhances the visual system's ability to transform that shape into the shape that follows it in the second flash. This finding creates a puzzle. Increasing the duration of a flash increases the detectability of the flashed shape, but simultaneously increases the likelihood that the shape will transform into the second. Hence, some information about shape is transmitted to higher regions of the visual system (recognition of the first flash), but the figure-changing aspects of the phenomenon seem to be more sensitive to local patterns of excitation than to identification of the shapes. The two following demonstrations make this point more clearly.

Let a hollow arrow be alternated rapidly with a somewhat larger rectangle, as shown in Fig. 4.7. When the distances θ and θ' are less than $2°$ of visual angle, the contours of the two shapes interact dynamically, the rectangle contorting to become the arrow and the arrow growing into the rectangle; this occurs irrespective of locus of fixation. In these movements, the observer sees contours bend and twist; it is not a matter of "inference of motion" (Dimmick and Sanders, 1929) but of clearly visible changes. When θ and θ' are about $3°$, location of fixation affects what is seen. Fixation on the upper edge of the rectangle is associated with a perception of the entire arrow moving upwards, then "bouncing off the ceiling", as one observer put it; when fixation is on the lower edge of the rectangle, the arrow seems to "bounce off the floor". Thus distance between borders determines whether the arrow expands into the rectangle, or whether the arrow retains its identity and moves as a whole up or down to one edge of the rectangle. Finally, when the distance between arrow and rectangle is 4°, some weak sense of motion is obtained, but it lacks the vivid dynamic changes of the other displays. Hence, maintenance of figural identity is dependent upon the local pattern of spatial interactions, not upon semantic features of the stimulus.

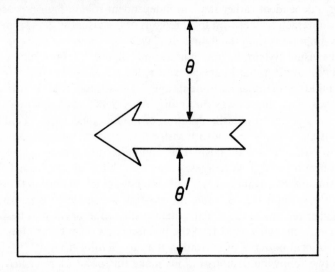

FIG. 4.7. The arrow and rectangle, when alternated, change plastically into each other when the distance between borders, θ and θ', is less than 2°.

The second demonstration amplifies this finding and illustrates another. Instead of an arrow, let the word BOX or the word CAT be alternated with the rectangle. When the distance from the word to the borders of the enclosing rectangle is about 2° or less, the rectangle contracts smoothly and continuously to become the "shape" of the word, but the letters of the word themselves "pop into place". The lines of the rectangle never appear to break to become the letters, and in the reverse direction the letters merely disappear and the outline of the word becomes the rectangle. Semantic features of the words in no way inhibit the motion; motion is perceived between the shapes when they are within proper distance irrespective of their contour and their identity. The smoothness and compellingness of the motion varies with the shapes, and the convincingness and continuity of the motion are greater in some cases than in others, of course. The point to be made however is that motion can be seen between any two shapes having the proper spatial and temporal characteristics, irrespective of their identity. Powerful interactions at a relatively primitive level of the visual system seem to dictate these results. Hence, figural changes and figural continuities are interpretations, rationalizations, or impletions applied to the interactions, but not the source of the motion perception. It is in this sense that I said earlier that patterns of excitation, not consciousness, dominates what is seen. The dependence of these effects upon distance separating the contours, moreover, makes implausible any explanation that speaks of inference or attribution of motion. Such cognitive events should not, by themselves, vary so precisely with distance. We find, rather, that the visual system accommodates a pattern of excitation. It does not begin with a shape that it perceives and then moves; it responds to a compulsory pattern of excitation and transforms visual figures to accommodate the pattern.

The classical argument is that the visual system perceives figures in different locations and infers motion to have occurred in order to resolve the disparity of figure location. What I have shown is that, to the contrary, the visual system responds to locations of stimulation and infers or creates changes of figure to resolve that disparity. Hence, "motion" is the compelling event, and figure or contour the dependent one in these results. Nevertheless, some information about shape of the stimulus is transmitted through the visual system, as other results indicate, so that in light of this fact the classical distinction between the two models, described at the beginning of this chapter, is seen to be inadequate. Rather than as figure perception or motion perception, it will be more useful to account for the results in terms of two analytically separable systems and stages of their operation. As an example, Table 4.1 illustrates the way phi motion, beta motion, simultaneity, and succession can be accommodated.

The table is a binary characterization of the phenomena, thus it does not represent partial motions, for example, although their interpretation in terms of the table should be clear. The table shows that when the visual system's threshold for motion is exceeded but the threshold for figure is not (M+F−), one sees objectless phi motion; when both thresholds are exceeded (M+F+), one sees beta or optimal motion; when the figural signal is adequate but the motion signal is not (M−F+), one sees simultaneity or succession; and of course when neither signal is strong enough, neither component is seen. The table illustrates only requirements, it does not speak of strengths or priorities. Hence this analysis will be pursued further, but first we need to examine some other aspects of the Gestalt theory of apparent motion.[5]

TABLE 4.1. TWO-COMPONENT CHARACTERIZATION
OF APPARENT MOTION

		Motion (H-Signal)	
		−	+
Figure (V-Signal)	−	Null	Phi
	+	Succesion Simultaneity	Beta

[5] While the present book was in press a paper by S. M. Anstis was brought to my attention that demonstrates in another way that form perception is not required for motion perception. See S. M. Anstis, Phi motion as a subtraction process, *Vision Research* (1970), **10**, 1411–1430. I thank Nicholas Mrosovsky for telling me of this work.

CHAPTER 5

ATTRACTIONS,
REPULSIONS, AND SIGNALS

ABSTRACT
The underlying mechanism invoked for most explanations of apparent
motion is an attraction between the figures in the two flashes. The concept of attraction is here examined analytically for the first time, and is
shown to be incorrect for the purpose. An alternative model is described
that distinguishes between signals related to the motion and to the
identity of the stimuli. Analyzing out these two attributes of the stimulus
simplifies the data somewhat, but the analysis is incomplete; this is shown
by experiments on crossed motion that yield perceptions of depth.
The conclusion is drawn that no two-dimensional model can accommodate the data fully.

Gestalt Psychology from its beginning emphasized forces and their
attraction as the analog of the perception of objects. Osgood (1953)
quite accurately characterizes the Gestalt assumptions about interactions, attractions, and the resolution of forces in apparent motion by
analogy to the interactions of magnets, whose strengths are variable
and whose attraction falls off with distance. In fact, there are several
Gestalt models, but they all share the characteristic of treating the
separate flashes as inducing some force that binds them into a unity
(Wertheimer, 1912; von Schiller, 1933). Let us consider some of these
notions.

Perhaps the most radical exposition of the magnetic model is found
in the paper by Brown and Voth (1937). Among the many features
discussed in their paper, two are especially notable. One is the "predic-
tive theory" Brown and Voth generated to account for their results;
the other is a figural distortion they observed. The figural distortion is
discussed in Chapter 9, and the theory now.

Brown and Voth (1937) proposed that a visual figure can be characterized by two kinds of forces, restraining forces that "allow us to account for stability of contour and of position, the figural properties (in Rubin's sense) of objects, and boundary phenomena" (p. 545, fn. 5), and cohesive forces that attract visual objects toward each other. These forces are alleged to operate as simple vectors (that is, as having both magnitude and direction) within the limits established by Korte's (1915) rules. The restraining forces are characteristics of static visual arrays, but the cohesive forces vary principally with time between flashes: "the magnitude of the cohesive vectors between two (perceptual) objects increases with a shortening of the time-interval between them until the stage of optimal movement, where this magnitude reaches a maximum. With further decreases in the time-interval the magnitude of the cohesive vectors must decrease, becoming equal to the restraining forces at the time-interval for perceived simultaneity" (p. 545). The authors state that restraining and cohesive forces can be equal, or be unbalanced in either of two ways, the restraining forces greater than or less than the cohesive forces.

To put it another way, Brown and Voth substituted physicalistic terms for visual events, and tried to account for the latter by analysis of the former. They allege that motion perception is due to an attraction between visual stimuli that are presented sequentially, and other perceptions are due to the weakness or absence of this attraction. This substitution aside, the model has the virtue of treating the motion experience not as the resolution of figural identities or maintenance of figural continuities, but as the simple resultant of patterns of excitation in the visual system. It has the further virtue of restoring, however obliquely, Wertheimer's distinction between the object seen as moving and the motion percept itself.

Although the model is only schematic, Brown and Voth claimed that it predicted many of the results they obtained, but not much weight can be given to this claim because the model is loose enough to predict a variety of events. Nevertheless, in this model specifically, and in the Gestalt models generally, attractional processes lie at the heart of the explanation. Separate figures that, when taken together, form a Gestalt unity, parts of a whole that either retain or lose their identity because of their placement, and even the notion of the short-circuit itself are all instances of an underlying concept of an attraction between figures or between parts of a figure and whole figures. Yet it seems to be the case that the underlying idea of attractions has never been examined analytically. Some experiments on apparent motion have yielded features here and there that are inconsistent with the Gestalt theory, but the plausibility of attraction as the mediating mechanism has remained unimpaired. The question of its validity in explaining apparent motion arises almost naturally, and should long ago have been tested.

The usefulness of the idea of attractions between visual stimuli depends of course upon the meaning of "attraction". In the Brown and Voth model each stimulus in a sequential display seems to be thought of as attracting the other, the resolution of their mutual attractions yielding the Gestalt. (Analogous suppositions characterize the Wertheimer and Köhler versions.) The fact is however that the duration of the first flash is more important to the motion perception than the duration of the second (DeSilva, 1928; Neuhaus, 1930; Kolers, 1964). Moreover, while simple vector addition may accommodate the results when there are equal numbers of objects in each flash, it does not work well when the numbers are unequal. To make a better test, we have found it necessary while exploring these puzzles to modify the Brown and Voth notions slightly. We assumed explicitly that all figures presented in a single flash repel each other, thereby retaining their location, while all figures presented in sequential flashes attract each other, thereby undergoing motion. The strengths of attraction and repulsion are assumed to fall off with distance. Thus, where Brown and Voth speak only of restraining forces to give each figure its stable contour, we add repelling forces to keep several figures in a single display from interacting.

In our experiments we utilized the phenomenon of split motion described by Wertheimer and by DeSilva (1926). Let the two flashes of a sequence be called A and B. If A consists of one circle, and B of two that flank it, the observer will sometimes see A fission and two circles move in opposite directions to become the two parts of B. (Some instances of split motion have already been illustrated in Arrays 2–4 of Fig. 4.6.) In our tests we used small circles, 6 mm in diameter (about 23 minutes of visual angle), arrayed on an imaginary horizontal line, as described in Chapter 4. Flash duration was set at 200 msec, and the interstimulus interval usually at 10 or 15 msec. Luminance of the fields

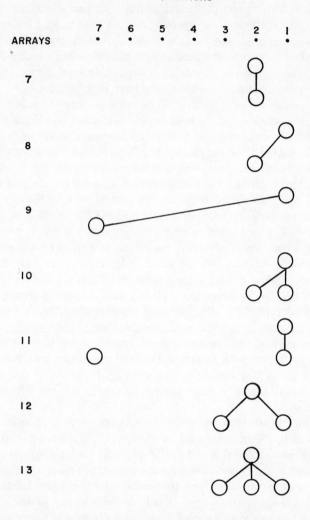

FIG. 5.1. Arrays 7–13. Tests of the concept of "attraction". In each pair of arrays the circles in the lower row appeared after a brief interval in the space occupied by the upper row. The connecting lines indicate what motions are seen.

was about 4 millilamberts. In some tests the intercycle interval was set equal to the interstimulus interval, yielding rapid cycling, and in other tests the intercycle interval was 3.2 sec. Occasionally, the length of the intercycle interval made a significant difference to the perceptual result (other than repetition); those conditions will be noted specifically. The tests illustrated in Fig. 5.1, as before, show pairs of rows of figures, with the convention that a difference in row indicates a difference in time, not in space, and a thin connecting line indicates motion. For convenience of description, however, I will usually regard the upper row of a pair as the first flash and the lower row as the second. Context should make it clear when this usage is not adhered to.

Array 7. The first flash is of a single circle at Position 2, followed by another circle in the same position. One sees only a single circle. In some sense the two flashes cohere into a single perceptual object.

Arrays 8 and 9. The two presentations exemplify the classical conditions of beta motion: one circle exposed in each flash is seen to move smoothly and continuously across the screen to the other. Only one circle is ever seen; and when the intercycle interval is short, the circle seems to be moving back and forth, either in quasiharmonic or in linear motion, depending upon the timing. When the distance between the circles is increased, the quality of the motion is sometimes affected, so that maintaining the perception of a smoothly moving figure may require some change of the intervals or of the flash duration. The spatial separation of six figure units (about 4.5°) was about the limit at which we could see smooth motion without requiring a change in the timing.

Array 10. Here we inquire whether duplicating a figure, as in Array 7, absorbs all the cohesive forces between the two flashes, or whether there is enough attraction between the flashes to move another figure as well. As the diagram shows, good lateral motion is seen between adjacent figures despite the duplication of one of them. The observer sees a single circle remain stationary at the duplicated location while at the very same time the same circle (or a copy of it) moves laterally.

Array 11. The same reasoning is applied to more spatially separated figures, but with opposite results. The observer sees a continuous circle at the duplicated location while the distant circle merely blinks on and off. Thus some of the attraction between distant points that is effective in

Array 9 is not available here; duplication limits the spatial range of the attraction.

Array 12. The classical demonstration of split motion finds that the single circle fissions and its "daughters" move off in opposite directions.

Array 13. Duplication is accompanied by fission; the observer sees a single circle continuously while copies of it move off in opposite directions simultaneously. Thus while the cohesive forces between duplicated points limit the spatial range of attractions (Array 11), the attractions are strong enough to operate over short distances (Array 10), even in opposite directions (Array 13). Moreover, observers do not report any marked difference in the quality of motion in any of these cases, nor, when the sequence of rows is exchanged, any difference in quality of motion between fission and fusion; between, that is, the separating and converging motions of Arrays 12 and 13. When motion is seen, it is smooth and continuous, although brightness or distinctiveness of the shapes varies, as would be expected: duplicated figures usually appear blacker than circles presented only once in a cycle.

All of the above demonstrations are consistent with a simple vector model that describes the motion perception as the result of attractions between stimuli presented sequentially. The experiments show that duplication of a figure does not by itself absorb all of the cohesive forces, although it weakens them and limits their spatial extent. Moreover, the effects described occur without regard to which elements of the display are fixated. In contrast, the following demonstrations reveal an influence of the locus of fixation and also reject the simple magnetic or vector model.

Array 13 duplicated the center circle but presented the flanking circles only once; good split motion was seen around an axis of symmetry. That contrasts with the following.

Array 14. The flanking circles are duplicated and the center circle presented only once. By extrapolation from Arrays 12 and 13, one might expect to see smooth split motion around the axis of symmetry, but the expectation is not borne out. The diagram in Fig. 5.2 shows that the flanking circles are seen continuously while the central circle moves either to the right or to the left, but not to both. When the eyes wander freely or fixate near the center of the display, the rightward motion is more likely. When fixation is on the rightmost circle, motion

ARRAY

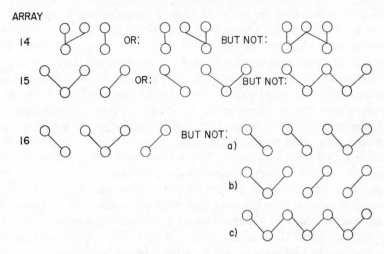

FIG. 5.2. Arrays 14–16. Some ambiguous displays disproving "attraction" and "symmetry" as explanatory concepts for illusory motion.

is principally to the left. Thus, motion is seen usually toward the peripheral visual field, but preferentially rightward from fixation. The array can be biased spatially by moving the rightmost circles farther right (not illustrated). The biasing, which might be thought to weaken the attractive forces between center and right, has only a limited effect: split motion still is not seen, but motion does tend to be seen more frequently toward the left, that is, over the shorter distance.

The results of Array 14 might be thought to illuminate the question of attraction. One might conjecture that the restraining (or figural) forces act against a shape's tearing itself apart, and that a shape can attract another shape only as a unity. Attractive forces might be conceived of as lodged in a single point at the center of a circle, say, and restraining forces as lodged in its contour. If this were so, then one might understand why a single flash cannot simultaneously supply attraction to another and the requirements for its own fission. The forces inducing fission would have to come from the shapes that act on another that undergoes fission rather than from the fissioned shape itself. Therefore, in Array 14, where the attractive forces of the flanks are largely used up in duplication, they are too weak to pull the central circle apart. The motion effect must then come from whatever attraction is in the central circle itself, which can be directed selectively to either side. Array 13 makes this a dubious conjecture, and the following test disproves it.

Array 15. As in the preceding case, three circles in the first flash are followed by two in the second, but now there are no duplications. An axis of symmetry is still preserved, but it seems to make little difference to the perception. The question is: does duplicating a circle have some unique effect that weakens the cohesive forces that might ordinarily cause a central circle to fission? If this were so one should see symmetrical split motion in Array 15; but as the diagram shows, the results are much like those of Array 14, and the effects of varying fixation are the same. It appears therefore that split motion fails to occur in Array 14 for reasons that are independent of repetition (duplication) of the flanking circles. These results suggest, rather, that a duplicated point is merely the limiting case for an attraction that is expressed over a distance. Duplication does not seem to release or repress factors that are not also present in other instances of optimal motion.

These results, which are inconsistent with a simple vector model, are not due only to the number of points involved.

Array 16. Additional circles were added to the preceding array, with the results shown in the diagram. Motion is typically seen from the center point to the two points that flank it, while the extremes appear to move in opposite directions. Here fixation of the eyes has little effect in modifying the perception, and none of the alternatives to Array 16 that are shown in the diagram are typically seen.

The description of Figs. 5.1 and 5.2 has reported the perception when the upper row of shapes in each diagram is flashed first and the lower row second, that is, as if the intercycle interval were always 3.2 sec. If in each case the lower row is flashed first and the upper row second, exactly the same motion relations are seen, but now the word fusion, as indicated earlier, must everywhere replace the word fission in the description. In Arrays 12 and 13, for example, the flanking circles converge upon and fuse with the central circle. When the intercycle and interstimulus intervals are equal, both fusion and fission are seen. Figural quality and the convincingness of motion appear to be the same in the two cases. It is not altogether plausible that these antagonistic processes, fission and fusion, have identical outcomes. The reason is illustrated with a little arithmetic.

Suppose that attraction is lodged in the shapes of a flash. Then if there is more than one shape in a single flash, that flash should have more attraction. Therefore, if the sequence in one test is one shape followed by two (fission), and in a second test two shapes followed by one (fusion), there would be two units of attraction for fission but only one for fusion. Fission should therefore be more compelling perceptually than fusion. A second possibility is that all the attractive force is lodged in the figures of the first display alone. Then, fusion of two shapes into one should be more compelling perceptually than fission of a single shape into two others. As noted above, however, there is no obvious perceptual difference between the two kinds of motion. Hence, attraction of *figures* cannot explain the results.

A third possibility is that the attraction is lodged not in the figures but in the flashes; then, all figures in a flash would divide the available attraction between themselves. For example, a single circle alternated with a circumscribing ring of circles will move smoothly in split motion, we found, as long as the ring contains fewer than about seven or eight circles. Say there are six. Then each of the circles in the ring would have one-sixth unit of attraction, while the single circle of the other flash would have a whole unit. Hence the same quantity of attraction, one unit, is lodged in both flashes, and fusion and fission should therefore be perceptually equivalent, as they seem to be. Although this analysis jibes better with our observations, it also implies that the hypothetical attractions are not in the figures or contours, as Brown and Voth maintain, but at best in the flashes themselves. Hence the motions cannot correctly be conceptualized as interactions between contours, but only as the results of the interactions between transients in the visual system created by the flashes presented to it.

It may be helpful here to summarize the main results so far. In Fig. 4.5 it was shown that the visual system readily performs a plastic deformation upon contours with no more trouble or cost than is involved in seeing them in motion. This conclusion emerges from the finding that disparate pairs of shapes are seen in smooth continuous movement with about the same ease as identical pairs of shapes. These results led to an examination of the idea that the motion system of the visual apparatus is primary and the shape system conforms to its dictates. In that light some further tests were made to examine first the role of shape in the motion perception (Fig. 4.6), and then the notion of a simple magnetic or vector model to account for the motion aspects of the perception (Figs. 5.1 and 5.2). This examination reveals that shape itself has relatively little influence on the motion perception; in addition, although the magnetic model can accommodate some of the simpler data, it breaks down rapidly when it is obliged to account for certain instances of "attraction", and especially the phenomenon of split motion. In the latter case one could expect differences in ease of seeing fission and fusion, but none was found. Moreover, such abstract notions as figural identity, symmetry, and number seem to have little relevance to the perceived resolution of the temporally disparate displays.

Non-duplicated Points

Some of the displays in Figs. 5.1 and 5.2 use duplication of stimulation; the results suggest that duplication has no special characteristics but is only a limiting case of distance. In the following experiments this conjecture was tested with displays that avoid duplication.

The reader should be aware that the notation used to represent these displays is different from that in the preceding figures; an additional dimension is represented. The stimuli used were always open circles, but to represent time, the first flash is shown as solid circles, the second as open circles; therefore, the diagrams illustrate the actual spatial arrangement of the displays.

The first array in Fig. 5.3 provides a *coup de grâce* for the simple vector model. If the flashes behaved as vectors, then vector addition requires that the single circle of the first flash, represented by a filled circle in Array 17, move at a 45° angle between the two circles of the second flash. This perception was never attained; only simple split motion was seen, as shown by the connecting lines.

Array 18 is analogous to Array 14 in Fig. 5.2 except that duplication is avoided. The results are exactly analogous, for split motion is seen between one element of the first flash and two of the second, but dual split motion around the axis of symmetry is not seen. Thus the inability to see dual split motion in Array 14 is not the result of duplication.

ARRAY

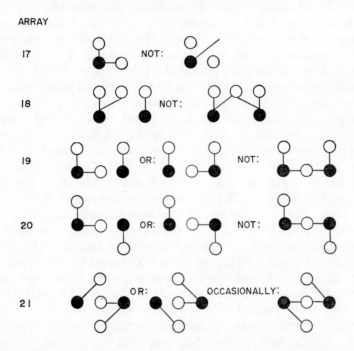

FIGS. 5.3. Arrays 17–21. Here and in the following figures the true spatial arrangement is illustrated, a difference in color representing a difference in timing. The figures drawn solid usually appeared first. The thin lines again indicate perceived motions.

Array 19 tests the role of element placement in the failure of dual split motion to be seen. That modification does not yield the perception either. The further modification shown in Array 20 also fails to generate symmetrical split motion. Hence element placement by itself does not explain the failure.

Array 21, the final demonstration in this series of tests, also fails to yield dual split motion when the intercycle interval is 3.2 seconds. This demonstration concludes the justification for the statements made earlier regarding duplication.

Arrays 19–21 show more than one configuration in the diagrams.

Some of these arrays in fact have several perceptual outcomes. When the intercycle interval is 3.2 seconds, an observer will typically report seeing a single configuration on repeated trials, but different observers will report one or another configuration. However, if the intercycle interval is made equal to the intestimulus interval (about 10 or 15 msec), then the entire range of possibilities reveals itself to a single observer. He will see one configuration for a number of repetitions, then report another, and another still. With prolonged observation and rapid sequencing, symmetrical split motion can in fact occasionally be seen. It is seen very rarely (perhaps 5% of the time) with Arrays 19 and 20, and a little more frequently with Array 21. Irrespective of the number of repetitions, it is never seen with Array 18, or with any of the arrays in Fig. 5.2. The effect of rapid cycling seems to be to allow the observer to sample among many different options in constructing the motion perception, as now one kind of visual construction and then another is used too frequently or "fatigues" in some sense (Orbach, Ehrlich and Heath, 1963). The main fact that remains, however, is that the vector or magnetic model must predict symmetrical split motion for many of the arrays shown in Figs. 5.2 and 5.3, but that perception is never reported with the arrays in the former figure, and only infrequently with the arrays in the latter. Duplication by itself is not the explanation, as Array 18 shows; distance apart of the elements or their direction of motion is also an inadequate explanation, as will be demonstrated below. But before turning to those figures, some more general issues need to be discussed.

The magnetic or vector model that has been under consideration is a metaphoric expression of the idea of attraction that underlies almost all theories (and many of the phenomenological observations) of apparent motion. In Chapter 4 evidence was adduced to justify the notion that not figures themselves but conditions of excitation that they generate are the main stimuli for the perception. The argument was made that perceived figural processes are rationalizations that occur fairly late in a figural construction. The magnetic or vector model assumes that the spatial and temporal relations of the perceived figures in their two-dimensional cartesian matrix can fully account for their perceived motion. It assumes, that is to say, that the perceptual processing is carried out on the "surface representation" of the shapes, or that what is seen describes how the visual data are operated on. The equivalence in respect to motion of shapes that are disparate in contour shows that this assumption is not correct. Moreover, the foregoing demonstrations reveal that the sort of attractional theory that is conventionally assumed to account for apparent motion cannot in fact account for it.

The attractional model fails even when the primitive concepts of restraining and cohering forces are supplemented with the concept of repelling forces. The notion of repulsion is needed if either of the displays in a sequence contains more than one element. For example, if three circles comprise the first display and two or four circles comprise the second, the notions of restraining and cohering forces, by themselves, do not explain why all of the circles in each display do not converge upon each other. Some "suprafigural" force is required to do that. Such a suprafigural force may be thought of as the Gestalt that a display itself represents; or it may be conceptualized independently as a repulsion between simultaneously presented elements. For convenience we chose the latter tack, but neither approach can do much to help the theory of attraction. Despite its historical position as the underlying assumption in accounting for apparent motion, attraction and its counterpart magnetic model have little if any strength for the task. In the following section an alternative is proposed.

A Two-component Model

The concept of attraction is clumsy at best; moreover, it cannot accommodate the facts of apparent motion. Something else is required, and consistent with contemporary interest, that might be a model that speaks of signals in a network rather than attractions between perceived objects. Several features that such signals must possess will be elucidated in course. The model proposed makes an interesting prediction that is borne out, but cannot yet be carried very far in detail. The sort of notion exemplified has already revealed other interesting characteristics (MacKay, 1970).

A visual stimulus whose image falls on the eye may be thought to generate two signals. One is a spread of excitation throughout the nerve fibers in the retina itself, which will be called a Horizontal signal. The other is a message to deeper parts of the nervous system, which will be called a Vertical signal. The Horizontal or H-signal is ideally suited to represent information about the *location* of a stimulus. The Vertical or V-signal is equally well-suited to represent information about *identity*. Thus the H-signal supplies information about *where* something is, and the V-signal supplies information about *what* it is.

Both kinds of signals would have the usual latencies and conduction times that characterize all signals in nervous networks. The stronger the stimulus, the shorter the latency until transmission begins and the more rapid the conduction. In addition, sites remain excited briefly even after the physical signal has ended. When a suprathreshold stimulus arrives at the eye, an H-signal will be generated through all the fibers that make connection with the site stimulated, its strength falling off with distance. (Because of the dense interconnections in the eye, two sites separated by a small angular distance could actually have longer paths connecting them and thus longer times for transmission than two more separated sites, for paths could fold back on themselves. A consequence would be perceptual overestimation of small physical distances.) A V-signal will also be generated that conveys information relevant to identifying the stimulus. The V-signal is however an amalgam, carrying information about many features of the stimulus. Hence there are either several kinds of V-signals, for contrast, contour, wavelength, and the like, or the V-signal must have signed components which are extracted separately; in either case, one component would carry information about motion, as follows.

Suppose that an H-signal from Location 1 (L_1) has been initiated in the network. While it is travelling, a second stimulus from the environment impinges on Location 2 (L_2). L_2 will generate its own H-signal (H_2) and V-signal (V_2). If H_1 arrives at L_2 while that site is still in a state of excitation, the coincidence in time of arrival of the two signals creates the necessary condition to generate what is interpreted as motion between the two locations. (Necessary, but not sufficient, as shown by Fig. 5.2. This is discussed further in Chapter 7.) The signal V_2 from L_2 is composite, however, carrying information about the identity of the stimulus at L_2 and the coincidence of signals. Such a composite signal would be stronger than the signal from L_1, and this difference in strength could be the analog of its sign. In addition, the difference in strength would allow V_2 to reach some deeper level of the visual system at about the same time as V_1. The difference in physical timing of the flashes would be compensated for by a difference in speed of conduction, permitting simultaneous or near-simultaneous arrival of the two signals in the brain. The two separate physical events at L_1 and L_2 would thus form part of a single perceptual event.

Suppose further that the threshold for H-signals is lower than the threshold for figural information in the V-signal. If L_1 generates a weak V-signal but a suprathreshold H-signal, properly timed stimulation of L_2 would yield the perception of objectless or phi motion; phi motion is perceived when the requirements of a motion-reporting system are met, but the requirements for a figural system are not (Table 4.1). We have found in our own studies, in fact, that phi motion is perceived when both flashes are weak, and also when only the first flash is weak but the second strong. In the latter case, people have a vivid sense of objectless motion that terminates in the appearance of an object: "whoosh-bang!" was one observer's description. But phi motion is almost never reported when the first stimulus is strong and the second one weak. Optimal motion or beta motion is perceived when the first stimulus especially is strong.

It is important to realize that coincidence of signals, and not mere stimulation of two sites, is required for this model. As in the traditional models, here too we say that if L_2 is no longer active when H_1 arrives, simultaneity of stimulation at L_1 and L_2 is reported; and if L_2 is stimulated after the effects of H_1 have dissipated, succession is reported. The onset-to-onset duration of about 30–50 msec sets a temporal limit on the perception of simultaneity from moderate flashes; the measured temporal limits would be an estimate of the speed of conduction of H-signals. So too temporal characteristics of backward masking, interference with the identification of flashes that are followed by neighboring ones (Raab, 1963; Kolers, 1968), may equally be used to estimate the speed of V-signals. Indeed, in one experiment that studied masking and movement simultaneously, Kolers (1963) found that masking a disk with a concentric ring did not prevent the masked disk from being seen in motion with another disk. We may say that the H-signal from the masked disk at L_1 reached a second disk at L_2 to yield a good perception of motion, while the V-signal at L_1 was interfered with. This experiment shows that the motion signals are processed more rapidly than the figural signals. It also demonstrates the utility of separating the figural from the motion components of the perception.

This model does away with attractions altogether and speaks only of the coincidence of signals. The implication of the model, however, is that L_2 is only a sink, its function merely to provide the phenomenal object with a place to go. Several experimenters, DeSilva (1928), Neuhaus (1930), and Kolers (1964) among them, have reported that varying the duration of the first flash radically affects the likelihood of seeing motion, but varying the duration of the second seems to have no effect as long as the second is visible. One might infer from this that apparent motion is due to a single generator function that is activated in proportion to the strength of the first flash alone. Then, when the first flash is weak, beta motion is weak, and as the first flash is strengthened, so too is the likelihood of seeing beta motion, as suggested in Figs. 3.4 and 3.5. But this cannot be the whole story and, indeed, we have found some evidence that questions the idea.

If a single black circle presented for 200 msec is followed by two others that are equidistant from the first, split motion is seen, as reported above. But if the two circles in the second flash differ in contrast, one drawn with India ink and the other with pencil, so that one is black and the other is gray, both circles in the second display will be visible, but motion is seen only between the black ones in the two flashes. The gray circle blinks on and off but is not seen in motion. Thus, aspects of the second flash, in this case figure contrast, affect the likelihood of seeing some forms of motion.

At first sight this result seems damaging to the idea that L_2 is only a sink. However, it is sufficiently ambiguous that we cannot yet tell. Perhaps the difference in contrast affects the summation of horizontal and vertical signals so that a properly signed V-signal cannot be generated from the site of the gray circle. Other conjectures are also possible. For example, some aspects of the second flash must enter into the figural construction when the shapes are disparate but the transformation reported is smooth. And although the contribution of shape to apparent motion is very small, it is not altogether absent (Kolers and Pomerantz, 1971). For the moment the role of the second stimulus in the apparent motion perception remains unclear, especially whether the second is active or passive, and whether it may exert an influence on both H-signals and V-signals.

Reverse Motion

Despite this uncertainty, an interesting prediction is generated by the model. Suppose the flash at L_1 is strong enough to induce a prolonged after-discharge in its detectors. A properly timed flash at L_2 would arrive in coincidence with the arrival of H_1, and motion would be seen between those sites, from L_1 to L_2, as in ordinary apparent motion. In addition, the detectors at L_2 would generate their own H-signals, which would reach a still-active L_1. Therefore, from a single sequence, L_1–L_2, it should be possible to see motion L_1–L_2–L_1.

We tested this prediction by making the first flash 250 msec, the second flash 10 msec, the interstimulus interval 10 msec, and intercycle interval 3.2 sec. Luminance of the fields was about 4 millilamberts. The stimuli were black circles as described earlier, one in each flash, their centers about 1° apart. With these arrangements the observer has a vivid sense of an object originating at L_1, travelling to L_2, and then springing back; that is, from stimulation at L_1 and L_2, one sees motion L_1–L_2–L_1. The moving object is not seen very distinctly, of course, for the second flash is very weak, but a vivid sense of motion is clearly present. We examined some of the requirements for "reverse motion".

Electrophysiologists have taught us that flashes to the visual system elicit several kinds of neural response: two of these are a discharge associated with the onset of the stimulus, and a discharge associated with its offset. The effect just described enables us to test whether reverse motion is selectively associated with neural discharges. It has long been known (Wertheimer, 1912; McConnell, 1927) that the two flashes may overlap somewhat in time (have "negative" interstimulus intervals) without impairing the motion perception. The query arises whether a perception of motion can be generated equally well by interaction of the onsets of two flashes, their offsets, or a combination of onset of one and offset of the other.

The illusory motion effect requires that transients be in the visual system from both flashes. If one stimulus is left on for, say, 5 seconds, flashing another during any part of the middle 3 seconds will not yield the motion perception. The observer sees one object continuously while the other blinks on and off (Kolers, 1964b). The tests with reverse motion suggest that although it requires transients, the visual system does not treat all of the transients in the same way. We found that the reverse motion effect disappears if (1) the interstimulus interval is increased to 50 msec or more, if (2) the second flash is increased to 50 msec or more, or if (3) the first flash is lengthened to 500 msec. Thus, the effect is not attributable only to the interaction of the onsets of the first and second flashes, nor to a sequence such as $Onset_1$–$Onset_2$–$Offset_1$. Signals associated with the onset and offset of *both* flashes are required for the reverse motion effect, revealing the interaction to be Onset 1–Onset 2–Offset 2–Offset 1. Presumably, for the effect to occur, the signal associated with offset at L_2 must reach L_1 while the latter is still active; hence the stimulus at L_2 must be brief.

This reverse motion effect may be related to classical delta motion. When the second flash is markedly more intense than the first, the illusory object appears to move in the direction opposite to the physical sequence of flashes. The flash sequence is L_1–L_2, but the perceived motion is from L_2 to L_1. For delta motion one assumes that H_1 is weak and H_2 strong; therefore H_2 arrives at L_1 before H_1 arrives at L_2, and motion is seen accordingly.

Despite its success in predicting reverse motion, the two-component model of signed signals is not wholly adequate, for it too fails to account for the absence of symmetrical split motion in Figs. 5.2 and 5.3. Difficult as that problem appears, even more difficult is the absence of "crossed motion" described below. One consequence of considering this failure is the realization that no two-dimensional model of the interaction of flashes can fully accommodate the phenomena of apparent motion.

Experiments on Interacting Paths

The diagrams in Figs. 5.4 and 5.5 represent the spatial arrangement of the flashes, again using the convention that solid figures represent the first flash and hollow figures the second. In the experiments, the figures were all hollow, and the timing of flashes the same as for Fig. 5.3.

In Fig. 5.4, Array 22 illustrates the motion seen when two circles are followed by two others positioned below them. Motion is seen

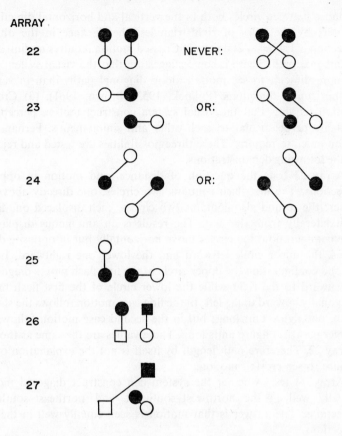

FIG. 5.4. Arrays 22–27. Failures of crossed motion. In the experiment all figures were hollow; those drawn solid here usually appeared first.

only in the vertical direction; it is never seen diagonally, crossing paths, regardless of where the eyes are fixated or how one "wills" the perception. With great effort, motion can sometimes be seen along one diagonal path, but it is never seen along both diagonal paths simultaneously.

What blocks crossed motion? At least three variables might be thought important. (1) Crossed motion increases the path length. The distance between circles both in the vertical and horizontal direction is 1 unit; by the rules of right-triangles, the distance in the oblique direction is 1.4 (= $\sqrt{2}$) units. (2) Crossed motion requires motion along a diagonal path. There is some suggestion that the visual system finds it more difficult to see motion along diagonal paths than in vertical or horizontal directions (Pollock, 1953; Brown, 1961). (3) Crossed motion requires that the visual system construct motion perceptions that are perpendicular to each other and simultaneous. Perhaps this is too much to require. These three possibilities are tested and rejected in the following demonstrations.

Array 23 tests the question of distance, and motion in opposite directions. The first flash contains two circles, one directly above the other; the second also contains two circles, each displaced one figure unit laterally from the first. The result is an ambiguous display. In one configuration, the circles move horizontally but in opposite directions, the upper circle leftward and the lower one rightward. In the second configuration the upper circle of the first flash moves diagonally downward to the right while the lower circle of the first flash moves diagonally upward to the left. In the first case motion follows the shorter path, one figure unit long; but in the second case motion follows the longer path, 1.4 figure units long. These values are the same as those in Array 22. Therefore path length by itself is not the explanation of the failure to see crossed motion.

Array 24 tests whether the system can construct diagonal motion equally well in the northwest–southeast and northeast–southwest directions. The answer is that motion is seen equally well in the two directions.

Array 25 inquires whether the fact that the two motions are perpendicular to each other blocks the occurrence of crossed motion. The diagram shows that motion can be seen in perpendicular paths.

The failure to obtain crossed motion in Array 22 was somewhat surprising. Higginson (1926b) had pointed out, correctly, that its occurrence is inconsistent with the Wertheimer–Köhler model, and also reported that he obtained many instances of it. Therefore we repeated all of Higginson's experiments, but in no case could we confirm his report. His objections to the Gestalt theory remain valid, but in light of this and other failures to confirm his experiments (for example,

Guilford and Helson, 1929), we may conclude that his results do not support his complaints.

In further efforts to obtain crossed motion we used displays such as that illustrated in Array 26. The shapes in the second flash appeared one figure unit below those in the first. Despite the opportunity for the shapes to retain their identities as circles or squares if they would go into crossed motion, only motion in the vertical direction, accompanied by plastic deformation, was seen. The circle of the upper row changed into a square while it seemed to move downward, and the square of the upper row changed into a circle. Hence neither contour, nor its placement, explains why crossed motion is not seen. Apparently, the visual system cannot tolerate what in appearance would be a collision between figures. The analog of a collision in the magnetic model is crossed lines of attraction or crossed vectors; in the two-component model the analog is crossed signal paths.

We sought to counteract the possibility of "collision" by offsetting the stimulus figures somewhat. Reducing the distance the circle in the upper row of Array 26 is required to travel could enable it to be out of the path of the square when the shapes move diagonally. Moderate offsets proved to be of no avail. Array 27 shows that the visual system required that the upper circle actually be below the path the square would travel before simultaneous diagonal motion could be seen. The visual system therefore will not allow two apparently moving objects to occupy similar regions of space at about the same time. Not only will the visual system not allow it, the system goes to some extraordinary lengths to avoid it, as the remaining demonstrations of this chapter show.

In Fig. 5.5, Array 28 is related to Array 22 of Fig. 5.4, but the lower row of circles has been moved one figure unit rightward. The two circles of the upper row appear first, followed by the two of the second row. In this array, motion follows parallel diagonal paths downward, as would be expected from the demonstrations in Fig. 5.4.

In Array 29, the vertical distance between the right circle of the upper row and the left circle of the lower row has been decreased by half a figure unit. No crossed motion is seen, in which the outer pair moves diagonally in the same plane and between the vertically moving inner pair. Motion can sometimes be seen along parallel diagonal paths, as

ARRAY

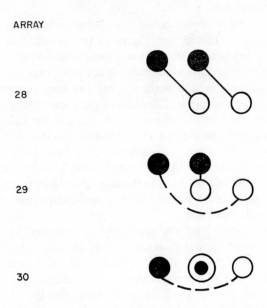

FIG. 5.5. Arrays 28–30. Depth as an alternative to crossed motion. The dashed line connects the figures seen in depth, it does not represent the detour implied by the drawing.

in Array 28, but especially when intercycle and interstimulus intervals are equal, the outermost pair of circles is seen to move in depth across the innermost pair, which move vertically. The dashed line in the diagram indicates depth, not the true path of the motion. The true path is an arc curving in front of or behind the central pair.

Array 30 shows the limiting case of the distance effect. The middle pair is composed of identical circles, one of which duplicates the other. Motion is again seen in depth, in front of or behind the plane of the central figure. The depth effect does not curve above or below the central figures, as the drawings might lead one to infer, but in front of or behind them.

The finding therefore is that the visual system avoids "collision" of illusory objects in either of two ways. In one, illustrated by Array 22, motion is seen only along parallel paths; in the second, illustrated by Arrays 29 and 30, motion is seen in depth. Depth as an alternative to collision has been reported before. Kolers (1964) and Goldstein and Weiner (1963, 1969) found that interposing lines, rectangles, or random squiggles into the visible field during the interstimulus or intercycle interval induces a perception of depth in the alternated stimuli. In one case, a grid was flashed during the interstimulus interval but not during the intercycle interval. Kolers reported that the illusory object moved in the plane of the display during the empty intercycle interval, and moved in depth during the filled interstimulus interval. The present results elaborate those findings.

The failure to see crossed illusory motion contrasts with what can be seen in veridical motion. Two spots in real motion on the face of a CRT can easily be made to pass through the same point simultaneously, as can the beams from two flashlights, say. The crossed paths that are never seen with illusory motion are easily seen with veridical motion. The visual system therefore is capable of allowing paths to cross each other perpendicularly and two lights to occupy the same point simultaneously when the motions are real, but does not do so when the motions are illusory.

Rather, when the motion is illusory and the path of one object would cross that of another, or even when a path passes close to a site stimulated by another flash, the visual system creates a perception of depth. (We have not quantified "close"; several observations suggest that it is about 1.5 figure units.) Hence the stimulus for depth, in these experiments, is proximity of temporally disparate stimulations and not, as in stereoptical depth, proximity of spatially disparate stimuli.

Proximity by itself, however, does not accommodate all the facts. It was shown in Chapter 4 that contour exerts a small influence on the perception of illusory motion in two dimensions. That influence is preserved in three-dimensional illusory motion, as the following demonstrations reveal.

The first array in Fig. 4.6 is redrawn as Array 31 in Fig. 5.6. In this figure we revert to the convention used earlier. The upper row indicates the first flash and the lower row the second, although in fact the shapes in the second row appeared in the locations of the shapes immediately

ARRAY

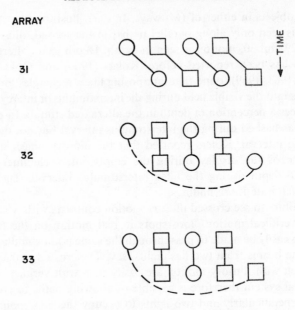

FIG. 5.6. Arrays 31–33. Further instances of depth, but now with spatially superposed presentations.

above them. The diagram schematizes the observer's perception of a line of circles and squares moving laterally. When the interstimulus and intercycle intervals are equal at 10 or 15 msec, the perception is of a whole line of figures oscillating in the plane.

Array 32 projects the results of Array 30 on to shapes like those in Array 31. The first row of circles and squares is now followed not by a copy of itself but by its mirror image. Thus, the middle three figures are duplicated, but the extreme figures are not. As the drawing shows, the middle three figures are seen in place, continuously, while the extreme circles orbit in depth around them. (As before, the dashed line indicates motion in depth, not the detour actually drawn.) Identity of contour in the superposed flashes distinguishes this array from Array 31. That identity is enough to inhibit the plastic deformation that might have been expected (for example, Fig. 4.6, Array 6), and to create a depth effect instead.

Despite this result, the mechanism for the depth effect is not identity of the extreme circles. Rather, it is the presence of stimulation between the extremes. That this is so is shown by Array 33, which is the same as Array 32 except that one of the extreme circles has been changed to a square. The observer sees the extreme object orbiting in depth around the intervening ones, changing shape as it orbits.

The influence of contour therefore remains small; it is expressed more by the maintenance of the central figures than by the extremes. The central figures in these arrays block crossed motion of the outliers, which then move in depth around the "interference". Why the visual system creates a perception of depth out of such arrays instead of allowing the outliers only to blink on and off remains as another puzzle. Concepts such as "attraction" are of little use in accounting for the results. The two-component model of H-signals and V-signals, as stated, is also inadequate before these data. That model has other virtues however and so I shall continue to use it in a metaphorical way, to separate the figural from the motion aspects of the perception.

Regrettably, we were not able, in the time available, to learn what conditions limit the crossing of paths of objects in illusory motion and how tridimensionality is constructed as an occasional alternative. There is some suggestion in the last arrays that a real object in the path of illusory motion is required for tridimensionality to be seen. This, however, is matter inviting further investigation. The perception of depth from suitably disparate arrays stands as a challenge for any theory of apparent motion, and some further instances of it will be described in the next chapter.

CHAPTER 6

MORE ABOUT DEPTH

ABSTRACT

Characteristics of apparent depth are described for both static and dynamic arrays. It is shown that the stimulus for depth is excitation of contiguous visual regions within certain temporal bounds, but the minimum definition of the necessary stimulus remains uncertain, for in some cases figure itself is found to play a role. Inductions of depth between fovea and periphery are described. It is also shown that depth need not be thought to come from motion, as some allege, but that whether motion or depth is seen depends upon the strength of signals. Motion and depth are described as the results of the same level of stimulus processing.

Many of the spatial translations of illusory motion can be accommodated with models that ascribe the results to an attraction between the two figures, but many others cannot; and many can be ascribed to a coincidence of two signals, one from within the visual system and the other from the environment, but many others cannot. Both of these models, the magnetic model and the two-component model, fail when other stimuli are flashed in the "path" of motion, such as shapes displayed during the interstimulus or intercycle intervals. In these conditions, the illusory object is seen to leave the plane of the display, coming out before or moving behind the objects in its path. Thus, an excitation in the path of illusory motion induces the path to become three-dimensional.

Some perturbations of the path can be accounted for in terms of local conditions of excitation, as Arrays 29–33 reveal. Related to these observations are the experiments of Deatherage and Bitterman (1952) and Shapiro (1954), which studied the effect of prolonged viewing of a shape upon the subsequent path of apparent motion. In both experiments the path was found to curve in depth around the region previously stimulated by the fixated contour. No "collision" could have occurred in these experiments because the extraneous figures had been removed before the conditions for apparent motion were activated. Hence it is not the phenomenological or figure-dependent interpretation of "avoidance of collision" that explains the construction of depth; rather, depth occurs as the visual system's response to excitatory conditions in neighboring perceptual locations. (By creating a perception of depth in these circumstances, the visual system avoids just the situation that we might conceptualize as a short-circuit. Hence the fact that depth is constructed as an alternative to crossed motion may be taken as the ultimate disproof of short-circuit theory, if disproof is still required.)

A number of other perceptions of depth may have the same mechanism as their source. When two different frequencies are fed to the vertical and horizontal drives of a CRT, a number of geometric patterns can be created as a function of their ratio. These Lissajous figures are easy to see in depth while the figures are in motion; the ease with which they are seen in depth seems to vary with the contiguity of their contours (Fisichelli, 1946). Related to these effects are the stereokinetic phenomena of Musatti (Wallach, Weisz, and Adams, 1956), and Wallach and O'Connell's (1953) Kinetic Depth Effect. In these cases, depth is seen when a cluster of plane figures or the shadow of a three-dimensional figure is in motion, and depth is never seen by the uninstructed observer when the figures are at rest. The importance to the depth effect of temporal and spatial proximity of excitations (the distance between moving contours) has been brought out by Fischer (1956) and by Ross (1967); many of the depth effects that can be seen with moving displays have been summarized by Braunstein (1962).

Most of our perceptions of depth depend upon the fact that the two eyes are so placed that when they are fixated on a single object they receive slightly disparate images of it. If a piece of glass or transparent plastic is placed before the eyes and the observer traces the contours of a visually fixated object, looking first with one eye and then with the other, the two drawings clearly reveal the disparity. Wheatstone (1838) did approximately this, combined the two drawings in a stereoscope, and thereby re-created the sense of depth available in the natural scene. [Panum (1858) measured how far apart the disparate images could be before they were resolved as two objects and their depth effect lost. This distance, Panum's Area, is at a minimum in the fovea and increases toward the periphery.] Paralleling Zeno's argument about figure and motion is the argument about figure and depth: investigators long assumed that in perceiving depth the visual system first apprehends the contours or shapes of objects and then resolves the disparity by creating a perception of depth. Julesz (1960) found, however, that slightly disparate regions of random dots can also yield a compelling depth effect in the absence of any contour or pattern detectable in either half-image. The depth effect seems to be due to local regions of excitation in a central station of the visual system rather than to the perception of figures, contours, or patterns.

A distinction must be drawn between the two sets of conditions yielding depth. In studying stereopsis, the investigator presents slightly different configurations to each of the two eyes, and both eyes participate in processing the stimuli. The various kinds of illusory depth described above, in contrast, can be seen irrespective of whether both eyes or only one eye views the changing scene; indeed, in the experiments reported in the preceding chapters all observations were carried out with one eye only. Although the mechanism for static depth from stereopsis cannot, because of this difference, be precisely the same as the mechanism for dynamic depth (depth through motion or temporal disparity), what may be the same is the interactive pattern whereby local regions of excitation from different sources yield the perception.

For this to be the case, the visual apparatus must correlate two regions of elements and their boundaries, whose pattern is often indistinguishable from its background. Both Julesz (1971) and others have speculated on the subtlety of that pattern-matching activity, but without elaborating any firm conclusions. In the absence of evidence to the contrary, one may conjecture that the mechanism and operations for the two kinds of depth are in general the same, but are distinguished by the place in the visual system at which they are carried out.

Efron (1957) and Ogle (1963) have shown that simultaneity of presentation to the two eyes is not required for a perception of depth; a limited region of asynchronies still allows depth to be seen. Julesz (1960) has shown that detectable contours in the separate displays of a pair are not required either. Now Kolers and Pomerantz (1971) have shown that two eyes are not needed for a perception of depth. Thus neither simultaneity, nor contour, nor binocular disparity is required.

The essential stimulus to depth seems to be only excitations of neighboring visual regions within certain temporal bounds. Julesz's research shows that "visual regions" is not to be identified as necessarily in the retina.

Influence of Contour

As I have shown, it is possible to accommodate a large number of illusory motion experiences by treating the effects as the result of local interactions; it is possible to extend some of these notions to the perception of depth. Nevertheless, contour is not without influence, and something of the way it expresses itself at an intermediate stage of visual processing can be inferred from the following experiments.

When circles, squares, or only flashes of light are the stimuli, the motion seen is almost always in the plane of the display. Wertheimer (1912) and Neuhaus (1930) showed, however, that if the stimuli have a discernibly disparate orientation, motion is seen in depth. Neuhaus presented two vee-shaped figures in opposite orientation. When the timing was proper, the observer saw a single vee-shaped figure turning through the third dimension. Thus one has a perception of depth dependent on an aspect of contour; trying to accommodate it theoretically as the result only of local interactions between regions of excitation is cumbersome.

Although it was known that depth can be seen when the flashes are figures that differ markedly in orientation, the effect has not been studied much. One question that arises concerns the conditions under which the visual system will "choose" plastic deformation or depth as the means of resolving the disparity between the presentations. Kolers and Pomerantz (1971) studied these choices.

The stimuli were right-angled trapezoids such as those in Fig. 6.1. As contours, all of the shapes are identical. They differ only in respect to orientation, as inversions (I), mirror reflections around a vertical axis (M), or rotations in the plane of the display (R). The two members of a pair were presented for equal but varied durations, a number of interstimulus intervals separating them. The intercycle interval was fixed at 3.2 sec, and during it the observer made one of three reports: whether he had seen motion accomplished by a smooth change of

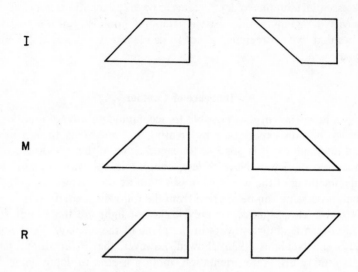

FIG. 6.1. The trapezoids in the right-hand column are 180° rotations of the trapezoid in the left column, either on the horizontal axis (I), vertical axis (M), or in the plane of the display (R). When two members of a pair are alternated, they can be seen in motion; their disparity is resolved either by a plastic change of contour or by rotation of a rigid object in depth. From Kolers and Pomerantz (1971).

shape in the plane of the display; whether the motion was accomplished by rotation of a rigid object in depth or, with R, in the plane of the display, or whether he had not seen smooth continuous motion on that trial. The observations were made with the right eye only, fixated on a small red dot of light.

One might think that increasing stimulus duration would, by enhancing their identity and perceptibility as stable objects, inhibit illusory motion of the disparate shapes. To the contrary, it was found that, like the effect of stimulus duration on the motion percept itself (Neuhaus, 1930), both smooth change of shape and rigid rotation in depth are more likely with longer flashes.

The two kinds of figural change have different temporal characteristics, however. Maximum likelihood of rotation of a rigid shape occurs at longer interstimulus intervals than plastic deformation. The actual

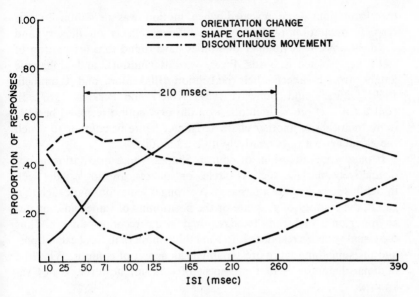

FIG. 6.2. The likelihood of seeing the trapezoids of Fig. 6.1 as changing shape, changing orientation, or as not wholly in movement. From Kolers and Pomerantz (1971).

difference in timing found in the experiment surely must depend upon its conditions, but averaged across the conditions employed was 210 msec. The differences are illustrated in Fig. 6.2, which also shows the proportion of trials on which smooth continuous movement was not seen.

Presumably it was the difference in their orientation that induced the visual system to transform the trapezoids by rotating them in depth. Sensitivity to orientation is a well-known characteristic of the visual system; hence one might further expect that transitions between the different orientations would not be seen in depth with equal facility. This was not the case, however.

Assessing our results statistically, we found that only about 1% to 3% of the total statistical effect could be attributed to orientation of the shapes. All of the trapezoids changed equally well into all of the others when motion was in the plane; only trivial differences between transformations were revealed when motion was in depth. In this respect orientation of figures has different effects on illusory and veridical objects. A number of studies have found that orientation of static arrays such as grids, faces, various contours, and alphabetic letters strongly affects their perception (Blakemore and Campbell, 1969; Gilinsky and Doherty, 1969; Kolers and Perkins, 1969). In contrast, the effect of orientation on illusory motion seems to be little more than that of another characteristic of figure or contour to which the visual system pays relatively little attention.

It pays some attention, of course. Shapes whose orientation is not distinctively marked, such as circles or squares, are not usually seen in depth when they are alternated. A strong orientation-mark such as a gap in the contour of a square or the positioning of trapezoids, creates an opportunity and seems to be required for a perception of depth. Thus the visual system's response is to something more than local excitations, and the something more presumably is an aspect of contour. What the minimal definition is of that aspect of contour, however, is not yet known.

Forms and Space

Plane figures that lack a distinctive disparity of size or orientation, such as pairs of circles or squares, are almost never reported in depth when they are alternated. Rather, their motion is seen in the plane of the display. This fact raises a question about the nature of the perceptual space in which the shapes are seen. Is the space itself necessarily three-dimensional, as a fixed attribute of the cognitive machinery; or does the dimensionality of the space depend upon the stimulation presented to it? In short, is depth in the stimuli or in their perceptual representation; and if the latter, is tridimensionality a fixed or variable aspect of perceptual space?

Figure 6.3 shows the spatial arrangement that was used in some tests of these questions. If only the upper square is followed by only the lower square, the flashes and the intervals between them all set at 100 msec, simple beta motion is seen: the square is seen to change locations smoothly and continuously. If now only the trapezoids are presented, everything else the same, smooth plastic deformation is seen in the plane of the display. However, if the durations of the displays (but not of the blank intervals) are increased to 200 msec, the trapezoids appear to change their orientation: they seem to rotate around a horizontal axis, in three dimensions. If now the arrays are made composite, so that a square and trapezoid in the first flash are followed by another square and trapezoid in the second, positioned below them, as illustrated in Fig. 6.3, what one sees depends upon where one looks. If the observer looks at or between the squares, the squares are seen in smooth beta motion while the trapezoids are seen, peripherally, in depth; but if one looks at or between the trapezoids, the trapezoids and the squares

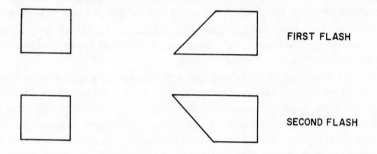

FIG. 6.3. Ordinarily, the squares are seen to move in the plane when they appear alone, whereas the trapezoids are seen to rotate in depth around a horizontal axis. When squares and trapezoids are both presented, the squares will sometimes also be seen in depth, depending upon the locus of fixation.

are all seen in depth, both shapes rotating on their horizontal axes through the third dimension. In short, the center of the eye needs its own discernible clue to a difference in orientation before it generates a perception of depth; it is insensitive to inductions from the periphery. However, when the center of the eye generates a perception of depth it induces a similar perception in the periphery. Hence, like motion itself, depth is created locally, but when it is initiated by the fovea, may generalize to other figures in the display. The depth is not in the optical array (Gibson, 1968), but in the nervous system's transformation of it.

We have found too that these induction effects operate not only over space, but over time as well. After a number of trials with trapezoids and squares that are seen in three dimensions, observers can sometimes see the squares alone in depth. A certain amount of voluntary effort is required for the perception, but it is distinctive. In this respect the results are very similar to those of Wallach and his colleagues on the kinetic depth effect. A stationary shadow of a three-dimensional object is not usually reported as being in depth; after viewing the shadow moving and deriving therefrom a sense of tridimensionality, observers report a stationary shadow to be in depth (Wallach, O'Connell, and Neisser, 1953). The initial stimulus for a perception of depth in our experiments seems to be duration of exposure and disparity of orientation. But once depth has been realized by the observer, it can be recreated when disparity of orientation is only imagined. In short, motion through the space in which visual figures are represented and, perhaps, even the representation of the space, is separable analytically from the figures themselves; moveover, perceptual space is subject to the observer's internal manipulation.

It has been shown therefore that a perception of depth occurs when the visual system would otherwise have to allow signal paths to cross in the plane, and when signals associated with another stimulus lie in or close to the path of motion. This depth perception is based not on contours, but upon local patterns of excitation. It has been shown also, however, that shapes whose disparity of orientation is clearly marked may themselves be seen in depth, but to accomplish this transformation the visual system requires more energy from the stimulus. When the duration of the flashes or the interstimulus interval is short, the system resolves the disparity between the flashes by deforming their shape. When the duration of the flashes and the interstimulus interval are sufficiently increased, the system resolves the disparity by rotating the figures in depth. (In this sense, the system has higher demands of the input to create a perception of depth than to create a perception of motion or plastic deformation.) And finally, when a perception of depth has been induced in a figurally similar pair by a figurally disparate pair (as the trapezoids inducing depth in the concurrently presented squares), the similar pair of shapes can also be seen in depth when they are subsequently presented alone. These results reveal therefore that the observer interprets in a similar way, a perception of depth, the results of three apparently different kinds of events in the visual system: local excitations, disparity of contours, and inductions.

Asynchrony

There is still another way that the visual system can create a perception of depth from alternated stimuli. Let two similar stimuli, such as two disks, be flashed at a rate that yields a smooth, continuous sense of motion in the plane, ordinary beta motion. Suppose this is accomplished by having the disks appear for 50 msec each and the interstimulus and intercycle intervals at 100 msec each. While the observer is noting this smooth oscillation in the plane, let the interstimulus interval be decreased to 50 msec, and the intercycle interval be increased to 150 msec. The appearance of the array changes sharply. Motion remains in the plane during one half-cycle, but comes out of the plane into depth during the other. The disparity here is only in the intervals between the flashes, yet the visual system resolves the inequality of the pauses between the flashes as motion in depth (Kolers, 1964). In this example, the ratio of interstimulus to intercycle intervals is as 1:3. However, the depth effect does not occur if the two pauses are 5 and 15 msec, or 25 and 75 msec, or 100 and 300 msec. Thus, it is not merely the temporal disparity, nor even the ratio of pause durations that induces depth; their absolute value as well is important.

At present, it is not clear whether the occurrence of depth from these temporal inequalities is a subtle example of one of the three processes mentioned above, or whether this is still another way that the visual system has for creating a perception of depth. Depth seems to be a common and relatively inexpensive means used by the visual system to accommodate a number of interactions between stimuli; although inexpensive, its occurrence requires a stronger input and more time to operate upon it (the interstimulus interval) than is required for motion or plastic deformation (Fig. 6.2). Despite these variations, we find that the visual system generates a "final common percept" (a perception of depth) by many different means. There are no unique conditions for either motion or depth.

Stages of Processing

Hierarchical models of stagewise processing have become popular in the recent past in accounting for visual information processing. The models have many forms, but in general assume that the perceptual experience (the picture in the mind) is constructed by means of a sequence of operations each of which is performed upon the output of a prior stage. People who create such models often look for the order or sequence of stages that characterize perceptual constructions. For example, Teuber, Battersby, and Bender (1960) and more recently Julesz (1971) have pondered the question whether motion precedes depth or depth motion, and where pattern analysis comes into the processing. Although models of these kinds are often useful conceptual aids, their authors sometimes neglect the possibility that a change in the mixture of inputs at a given stage of analysis can be the correlate of a change in perception.

It was shown in the preceding experiments not that depth is derived from motion but that a change in the strength of the input determines whether planar motion or motion in depth is seen. (It is meaningless to inquire whether depth requires motion if one can create a sense of depth from unmoving temporally superposed shapes.) The visual system may indeed construct its perceptions in a stagewise way, as will be discussed later, but the nature of the stages is not as rigid and fixed as some authors seem to imply. For example, Table 4.1 schematized the results for stimuli activating the motion and the figural systems; the analysis assumes that the two flashes of a pair are equal in energy. Some other phenomena occur when they are made unequal. Table 6.1 schematizes the results.

Here, again for convenience, stimuli are dichotomized as strong and weak, the former about 200 msec in duration, the latter one-tenth that, and luminance about 4 millilamberts. When both flashes are weak, as described earlier, phi motion is seen; if only the second is strong, the variant on phi motion is seen that one observer described as "whoosh-bang". When only the first flash is strong, beta motion is seen preferentially, but depending upon the shapes actually presented, motion in depth can be seen as an alternative. When both flashes are strong, depth is the likely outcome when the shapes differ in orientation, but can also be seen when they do not.

The interaction of disparity of shape and strength of the flash is schematized in turn in Table 6.2, but in this case "weak" means a flash of 30 to 50 msec duration. When the shapes are identical but the flashes weak, simple beta motion is seen. When the shapes are disparate the variant on beta motion called "plastic deformation" is seen. When the shapes are identical and the flashes strong, beta motion is more likely although depth can be seen. When the shapes are disparate and the flashes strong, motion is seen in depth. (These tables can be related to the results shown in Fig. 6.2.)

TABLE 6.1. EFFECTS OF VARYING SIGNAL STRENGTH

| | | First flash | |
		Weak	Strong
Second flash	Weak	Phi	Beta (depth)
	Strong	"Woosh-bang"	Depth (beta)

TABLE 6.2. INTERACTION OF SIGNAL STRENGTH AND FIGURAL DISPARITY

| | | Disparity | |
		−	+
Signal strength	Weak	Beta	Plastic deformation
	Strong	Beta (depth)	Depth

The perceptual outcome, therefore, is keyed not to an order of precedence or rigid sequence of operations, but to variations in the strength of the signals. A perception of motion and a perception of depth both presuppose disparity of locations in the eliciting stimuli, so the two perceptions do not differ in this respect. They do differ in respect to the level of energy that is required to be in the two flashes, motion requiring less energetic and depth more energetic flashes. Motion does not necessarily come from depth nor depth from motion; both events seem to be aspects of the same stage of visual processing, not of serially arranged stages. The occurrence of motion or depth depends upon the strength of the eliciting signals primarily, and, to a lesser degree, upon contour.

CHAPTER 7

CHARACTERISTICS OF FIGURES

ABSTRACT

Although figure or contour plays only a limited role in apparent motion, it does play some role. A model based on "receptive fields" cannot accommodate the data, despite the importance of "excitations" and the relative unimportance of figure. The roles of figural properties such as connectedness, size, and grouping are discussed. The phenomenon of "replacement" is described, and the results obtained are related to current efforts to notationalize shapes. Distinctions are drawn between featural and constructional similarity. A model covering some aspects of the processing of illusory motion is described.

In the preceding chapters it has been shown that figural characteristics of the displays play only a small role in beta motion. Although small, a reliable contribution to the occurrence of beta motion is due nevertheless to characteristics of shape such as orientation, size, distance between the figures, and the like. Some of these effects vary with strength of signals and distance apart of stimulated loci, but some may be attributable to actions performed upon the neural correlate of contours themselves. Figure-dependent operations could be carried out, perhaps, at a different stage of stimulus processing from that concerned with strength of signals. By "different stage of processing", I mean to separate the energetic from the graphemic component of stimulation, much as one can separate the graphemic from the semantic, as in Fig. 4.2.

Size and Distance

Size and distance apart of the stimuli are important parameters for the perception of illusory motion, but the role of size is actually ambiguous. In the earlier literature some authors reported that smaller figures were seen more readily in motion than larger ones, but other authors reported the opposite result; the discussion was summarized by DeSilva (1928). In any case, the results were not expressed quantitatively; the authors were concerned with the quality of perceived movement and its resemblance to the perception of real movement. They expressed their results in terms of its apparent speed, jerkiness, compellingness, continuity, and the like rather than as quantitative functions of stimulating conditions. Necessarily the quality of motion will vary in a U-shaped fashion as a function of size of stimulus: if the flashes are minute, the motion perception will be less compelling than if the flashes are larger; but if the flashes are very large [20° or more of visual angle, in DeSilva's (1928) report], compellingness and vividness of the motion perception again break down. It seems likely that there is an optimal size of stimulus for beta motion, but what it is, and how it varies with retinal location, intensity, or even context have not been measured. Moreover, it seems likely that variations in size may allow other factors to express themselves. For example, we found that a small square (0.3° of visual angle) that is flanked on one side by a similar square and on the other by a circle of equal angular size always moves simultaneously in split motion to both flanks. If the figures are enlarged somewhat, to about 1° of visual angle each, repeated cycling of the displays allows some figural selection to occur; the central figure seems to move preferentially to its mate. We have not studied these effects quantitatively, but they suggest, as will be brought out again below, that new possibilities of perception arise as figure size is varied.

It should be interesting to learn whether size of stimuli in apparent motion has the same effect as size does upon figural grouping. Three dots can be seen as a triangle when their distance apart is not too great (Wertheimer, 1923). First we should like to know how the compelling perception of the dots as a triangle varies with their distance apart; then we might inquire whether this rule of grouping yields similar results for dots seen in illusory motion. Experiments such as those illustrated in Chapter 5 suggest that motion may weaken the bond that enables elements to appear grouped together in static displays. Von Schiller (1933) maintained that the same rules of grouping governed static and dynamic displays, but did not try to measure differences between the two conditions. Johansson's (1950) ingenious research revealed a number of kinds of figure formation in moving displays. The question cannot yet be answered whether the visual system's tendency to group elements into figures expresses itself identically on moving and stationary displays. The issue is not whether motion precedes or follows grouping in the organization of perceptions, but how these two events interact.

Not only distance between the parts of a single display but distance between the flashes is an important determiner of whether apparent motion can be seen. Teuber and Bender (1950) reported that their observers could see motion when the flashes were as much as 22° apart, but their report does not make clear whether it was objectless phi motion or smooth continuous beta motion that was seen. With flashes of moderate intensity, Neuhaus (1930) reported, motion failed when the flashes were more than 4.5° apart (see Fig. 3.2), and he too does not make clear whether the response was phi or beta. It is still not known what the maximum spatial separation between the flashes may be and motion still be seen. Stimulus duration and interstimulus interval affect the likelihood of seeing motion, but even with these variables controlled, it seems probable that phi motion will be seen over greater physical separations than beta motion. That is to say, the perceptual synthesis of H-signals can be accomplished over a wider separation than the synthesis of the corresponding V-signals. In experiments with the hollow arrow and surrounding rectangle described in Chapter 4, we found that good beta motion with plastic deformation can be seen only when the contours are less than about 4–5° apart. It would be most useful for a theory of motion perception, indeed, for theories of perception generally, to learn whether distance affects both figural beta motion and objectless phi motion in the same way. (I anticipate a negative answer.) These results would in turn affect the questions that could profitably be asked about grouping of elements in a single flash.

Table 7.1 summarizes the limits that have been found for various spatial interactions in vision. The narrowest range is Panum's Area, the distance apart two test lines can be from a common reference and still be seen as single when they are presented one to each eye. Panum's Area is of course a measure of binocular fusion, hence is the opposite of spatial resolving power. When one of the test lines is about 2° from the fovea and the test objects are presented for unrestricted durations, Panum's Area is about 10 min of visual angle (Ogle, 1950). When the

TABLE 7.1. SPATIAL RANGE OF INTERACTIONS WHEN ONE TEST OBJECT IS ABOUT
2° FROM THE FOVEA AND THE SECOND IS MORE PERIPHERAL
(Degrees of visual angle)

Event	Spatial range	Source
Panum's area ("static")	About 0.16°	Ogle (1950)
Panum's area ("dynamic")	About 1.0°	Fender and Julesz (1967)
Backward masking and metacontrast	About 1.0°	Alpern (1954); Kolers and Rosner (1960)
Plastic deformation	About 2.0°	Fig. 4.7
Beta motion	More than 4.0°	Figs. 3.1 and 3.8
Phi motion	More than 10°	Anecdotal

measurements are made with stabilized images of lines that are presented for controlled durations, Panum's Area may extend to about 1° (Fender and Julesz, 1967).

Backward masking of dichoptic stimuli has approximately the same range Fender and Julesz measured for Panum's Area (Kolers and Rosner, 1960), and this is about the range also measured for monoptic metacontrast (Alpern, 1954). Plastic deformation of contours extends over a larger range, at least as measured by objects such as the arrow and enclosing rectangle (Fig. 4.7). Our apparatus did not allow us to explore the limits of the spatial range of beta motion with disparate shapes. Measurements made by Neuhaus (1930) and measurements from our experiments with flashes on a CRT (Chapter 3), where identical lines were flashed as the stimuli, show beta motion extending over at least 4° of visual angle, or more than four times the distance found for masking and metacontrast. Phi motion may have even a greater spatial extent. All of these distances vary with the part of the eye stimulated, the spatial range of interaction increasing as one moves from the center into the periphery. How these various figural interactions differ in their local processing, and what their relation is to visual receptive fields is not yet known.

Figural operations may have distinctive requirements in apparent motion. In one experiment Kolers and Pomerantz (1971) alternated right-angled trapezoids, illustrated in Fig. 6.1. The members of a pair were either inversions, mirror reflections or planar rotations of each other. As described earlier, the subjects reported whether they saw a plastic shape moving in the plane of the display or a rigid shape changing its orientation. Rotations in depth for mirror reflection and inversion were quite straightforward: one trapezoid rotated either on its horizontal or its vertical axis to become the other. When the shapes were planar rotations, however, two options became available. The observer occasionally saw a trapezoid rotate 180° in the plane of the display and move laterally; more often, he saw a trapezoid invert and reflect simultaneously, moving out of the plane as it did so. The two options yield identical results. Why the visual system preferred performing the multiple operation of inversion and reflection rather than a simple rotation in the plane is not known.

In all transformations involving change of orientation the visual system necessarily must pay attention to something more than an excitatory signal itself, for otherwise an appropriate figural transformation would not be accomplished. However, the matter seems to be quite complex. The following demonstrations suggest some of the conceptual problems. They also reveal that at least two different kinds of figural processes seem to be active in the perception of illusory motion.

Connection and Separation

In one series of tests we studied the role of figural similarity and connectedness. The displays were always hollow rectangles each of which subtended 0.5° × 2.5° at the eye; they appeared in a lighted field of 4 millilamberts. The flashes were presented for 100 msec each, usually with the interstimulus interval at about 75 msec and the intercycle interval at 3.2 sec. Although all of the figures were hollow, in the diagrams of Fig. 7.1 some of them are drawn filled in. This convention indicates that the filled figure was presented first.

Arrays 34 and 35 in Fig. 7.1 are control conditions for Array 36. In Array 34, the hollow central rectangle is seen to split and move simultaneously to both flanking rectangles; in Array 35 the central rectangle is seen to split and its parts to rotate in opposite directions to become the flanking horizontal rectangles. In Array 36, however, some figural selection is manifested: most of the time the vertical rectangle of the first flash moves to the vertical rectangle of the second flash while the

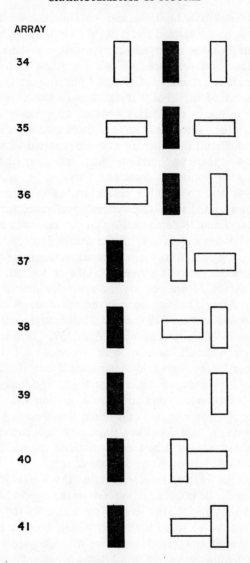

ARRAY

34

35

36

37

38

39

40

41

FIG. 7.1. Arrays 34–41. The role of connectedness in determining what moves. All the shapes were hollow, but color is used here to indicate sequence: the solid shapes were usually presented first.

horizontal rectangle only blinks on and off. Thus, despite the ease with which it can rotate the vertical rectangle, shown in Array 35, the visual system sometimes takes the opportunity to move a shape selectively, as was mentioned above. Therefore, while the first flash of an array may generate an undifferentiated H-signal out through the visual network, coincidence of arrival of that signal and another signal from the environment are not by themselves sufficient to generate a perception of motion when the figures are as large as those used here. These results are markedly different, therefore, from those reported with the smaller arrays in Fig. 4.6; there we found no figural selection. Hence size does indeed allow other options to come into play.

Arrays 37–41 test some other characteristics of figural selection. Array 36 demonstrated the visual system's preference for an identity match. Arrays 37 and 38 demonstrate that this tendency is weak. The first rectangle always moves only to the nearer figure, to the vertical rectangle in Array 37 and to the horizontal rectangle in Array 38. In both cases the more distant figure only blinks on and off. The vertical rectangles of Array 38 are not too far apart for motion to be seen between them: Array 39 shows that the vertical rectangle of the second flash can be seen in motion with the first. Thus figural identity is not an important feature of the processing of Array 37. These results are to be compared with those illustrated with Arrays 40 and 41. With the latter arrays motion is always to the composite figure; if the interstimulus and intercycle intervals are made equal, the composite figure changes plastically to a rectangle in one direction of motion and the rectangle grows into the composite figure in the other. Rapid cycling has no such effect on Arrays 37 and 38. Thus proximity and connectedness of borders can affect what will be seen in motion. The tendency toward figural selection revealed by Array 36 is weak at best.

Even with larger figures, therefore, we find that figural identity plays only a very small role in beta motion. Given the opportunity to select a figural mate in Array 38, the visual system disdains the opportunity and chooses instead to move the nearer figure. In at least one sense this is not a surprising result. It was shown in Chapter 4 that plastic deformation of rather extensive proportions is accomplished about as readily by the visual system as beta motion between identical shapes. As it can accomplish these figural transformations so readily, there seems little reason for the system to concern itself with the identity of what it is moving; hence more fundamental characteristics such as distance and connectedness take precedence over more abstract ones such as identity and symmetry.

The visual system is not responding only to excitations, however, as will be shown again below, nor moving everything on its retina indiscriminately. Connectedness of contours is an important determiner

ARRAY

42

43

FIG. 7.2. Optimal motion is seen between the single circle (solid) and the other four irrespective of whether motion is seen in a single direction (for example upward) or to the four corners of a square.

of motion, but not essential to it. Before discussing the matter of figural analysis in detail, however, I will show that connectedness of parts is not essential to motion. Arrays 42 and 43 of Fig. 7.2 indicate that a single circle, 23 minutes of visual angle, was followed by four identical circles arrayed symmetrically around it, or by four circles arrayed in a line above it. One might think that the lack of connection between the parts in the second flash would affect the motion perception. One might also think that the similarity of direction in Array 43, compared with the differences in direction in Array 42, would affect the perception.

However, in both cases the single circle is seen to fission smoothly into four others. If the interstimulus and intercycle intervals are made equal, the four converge upon and fuse with the one and the one fissions into four equally well in both arrays. Like identity, symmetry, number, and related mathematical characterizations of shapes, redundancy of direction is equally unable to accommodate many experimental results.[1]

The results of Array 39 show that distance by itself is not a sufficient answer to explain the selectivity of motion in Arrays 37 and 38. An

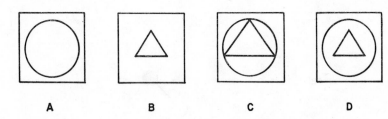

A   B   C   D

FIG. 7.3. Selectivity of the motion percept. When the square is alternated with the enclosed shapes in A, B or C, smooth deformation is seen; in D, however, motion is seen only between circle and square, while the triangle blinks on and off.

alternative explanation considers the existence of other contours in the path of motion. Contours, as Fry and Bartley (1935) showed, can limit the spread of excitation on the retina. Hence, the intermediate rectangle can be thought of as absorbing the attractive forces or H-signals from the first flash, thereby causing the more distant figure to lie in its shadow. To put it another way, one contour may "protect" a second.

We tested this supposition with the shapes illustrated in Fig. 7.3. Let a circle of 3° angular subtense be alternated with a square just enough larger than the circle that no parts of the two shapes can be seen to touch (A). The two shapes will, we know, transform into each other smoothly. When the circle is replaced by a triangle whose altitude and base are both a little less then the diameter of the circle (B), square and

---

[1] We have found that the visual system can generate about six or seven "daughter" figures in apparent motion when the stimuli are like those of Fig. 7.2; fission and fusion are seen equally well with such arrays, as described in Chapter 5.

triangle will also be seen smoothly in motion. Now let the circle and a triangle be part of a single display. If the triangle is inscribed within the circle, its vertices touching the circle's perimeter (C), the composite figure transforms into the square, as with Arrays 40 and 41. However, if the triangle is drawn as in (B) so that no part of it touches the perimeter of the circle (D), only circle and square are seen in motion, and the triangle blinks on and off. When the triangle is in contact with the circle (C), it does not disengage itself and refuse to go into motion. In this sense, the border of the circle does not itself "protect" the enclosed triangle. When the contours are separated, however (D), the triangle is able to maintain its own identity, "protected" by the shadow of the circle. In light of these results, it seems plausible that the more distant shapes of Arrays 37 and 38 do not move, less because of their distance from the first shape, than because they are not connected to their neighbors, whose contours shield them from the H-signal. Connected shapes therefore behave as figural units, as indivisible wholes rather than as analyzable composites.

The visual system, to put it another way, is not analytical for figure in these tests. This conclusion is supported by other tests as well. Kolers (1964) reported that when a Necker Cube was alternated with a component part such as a square, the visual system did not analyze the composite and move only the common part. Rather, the square grew into a cube in one direction of motion, and the cube shrank to a square in the other direction. Figure 7.4 illustrates some other ways that a Necker Cube can be sectioned. When any of these parts is alternated with a whole cube at the proper rate, the perceived result is plastic deformation. The visual system does not abstract out from the whole cube the part that is being alternated with it and allow the rest of the cube to blink on and off. It moves the whole cube, the whole figural unit, "growing" the part into the whole or "shrinking" the whole to the part. This occurs equally well when the parts are connected, as in a–c, or separated, as in d.

A clear definition of figure or figural unit is quite difficult to come by, however. The term cannot be identified with continuous contours, despite the fact that continuity of contour determines some of the results just described. The reason is that dotted figures, such as Djang (1937) and others have employed, go into motion as well as whole figures do;

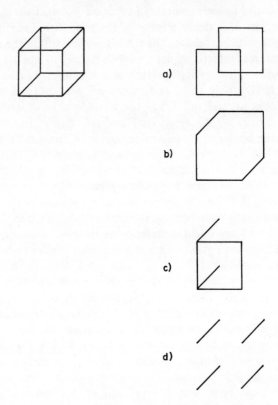

FIG. 7.4. Further instances of figural synthesis. A Necker Cube alternated with any of its features in *a–d* moves as a whole figure; the visual system does not select out the common part for movement.

all displays of dots or blobs that we have tried will seem to move as a whole, as long as the blobs or dots are close enough to seem to "belong together". The concept of figure is too vague to explain much about beta motion, yet something about shapes or figures does affect the perceptual outcome of beta motion. That "something" cannot be "identity", similarity, or even contour itself; it requires a more abstract definition. Clearly, grouping of parts—spatial density—is important to these effects, but not enough is yet known about grouping for us to find explanations in that concept.[2]

It might be thought, as an alternative, that the visual system is sensitive not to figure but to excitation alone within a region or zone of the eye ("receptive field"), so that the role of figure is only a minor influence on that sensitivity. Then, according to this reasoning, any stimulus that occurs within such a zone would be seen in movement with any other. This hypothesis would accommodate the results illustrated by Figs. 7.1, 7.2, and 7.4, all of whose individual parts could fit into a receptive field (Jacobs, 1969). Figure 7.3 tests this idea, and Fig. 7.3D reveals that it cannot accommodate the results; an enclosed figure that would fall within any receptive field relating the circle and square is itself not seen in movement. Excitation of receptive fields and the operation of feature analyzers within them cannot by themselves be the basis of the results.

In the other figures mentioned, the visual system reveals itself as unanalytical for contour, preferring plastic deformation of moving contours over selection and analysis of common parts, both when the parts are connected (Fig. 7.1 and 7.4a, b, c), and when they are separated (Figs. 7.2 and 7.4d). Figure 7.3D, however, proves that the visual system does not move everything within a zone, hence does perform some kinds of figural analysis. Regrettably, we were not able to work out the conditions in which analyses are carried out and when not. The major finding that is established is that figural transformation is preferred as an option to figural analysis. The major problem revealed by this and related findings is the difficulty in specifying more precisely the definition of figure and the characteristics of figures that the visual system does analyze and operate on. The usual definition based on connectedness of parts, name, or descriptive geometry (the manifest representation of figures) are shown in these experiments to be insufficient.

[2] When some of these experiments are carried out with displays on a CRT, the sense of motion is often less compelling than with the simultaneous display of all parts of a figure created by a tachistoscope. The visual system may be sensitive to the writing rate of the CRT, constructing simultaneous tachistoscopic exposures differently from sequential oscilloscopic exposures.

Replacement

When, at the proper timing, a person sees an object and a copy of it in two different locations, his visual system supplies additional copies in the perceptual space between; the system fills in or impletes what might be thought of as redundant copies between the termini. When the two objects flashed are different, however, the visual system transforms one into the other, smoothly and continuously, resolving their difference rather than supplying redundant copies. The perceptual experience, therefore, is constructed partly from information obtained from the environment, and partly from information that the observer supplies himself. The self-supplied information clearly is figural, contours and shapes undergoing transformation in the space between the termini. Let us consider some aspects of these figural changes.

In the preceding pages it was reported that any picture or diagram can be seen in motion with any other, and in accounting for the phenomena the impulse to motion was given priority over the perception of figure. In some cases the quality and smoothness of the figural change is poor, however; a sense of global motion is obtained, but the observer is unable to say precisely what changed into what and how the change was effected. Indeed, in their report on motion between disparate figures, Kolers and Pomerantz (1971) indicated that whereas smooth and continuous motion could usually be seen between circle, square, and triangle, any of these figures alternated with a hollow arrow sometimes yielded a perception better described as motion with replacement than as smooth continuous motion. The arrow, for example, would be seen to move smoothly and continuously across the screen, changing its shape somewhat, and then suddenly be replaced by the triangle, circle, or square at the terminus. Their observers found it difficult to distinguish motion with figural replacement from smooth continuous plastic deformation, however, so both reports were considered as good motion.

Despite the difficulty of quantifying aspects of the perceptions, the visual system does seem to be doing different things in the two cases. In one case it transforms a plastic figure smoothly into another; in the second it moves a plastic figure and then replaces it suddenly at the terminus. These two effects represent differences in the visual system's performance that may be attributable to the contours or figures being moved.

In trying to analyze this difference in appearance, one encounters a methodological, almost an epistemological difficulty, however. The phenomenon of illusory motion occurs in part because the stimulus conditions exceed the analytical capabilities of the visual system. When the presentations are brief and rapid, the system does not analyze them into their elements; following this failure, it does not even move common features of the arrays. Rather, it moves entire figures. We are dealing not with an analytic but with a synthetic system, one almost that rejects analysis as a means of operation. Yet in requesting the observer to distinguish smooth continuous change from motion with replacement, we are requiring the system to be analytical about its nonanalyticity, to describe differences between figures which, were they important to the system, would certainly be maintained, but are not. This fact creates a problem not met in the usual psychophysical experiment, and forces us to depend more on qualitative than quantitative accounts.

The experiments on surrounded figures, described in Fig. 7.3D, reveal one aspect of the phenomenon of replacement. A triangle enclosed within a circle whose borders it does not touch blinks on and off when the composite figure is alternated with a surrounding square, while the circle changes smoothly and continuously into the square. Similar results are obtained when the circle contains within itself smaller concentric circles, or when a large square is alternated with a smaller square that contains still smaller squares within its border. The inner shapes disappear and reappear, while the outermost contour moves smoothly with the large surrounding square. The effect is most vivid in two other conditions.

Let a series of concentric circles and a series of squares (Fig. 7.5A) be alternated. When they are superposed temporally, all of the squares change smoothly into all of the circles, and vice-versa. Similarly, let the squares of Fig. 7.5A be alternated with the grid of Fig. 7.5B; when superposed, the shapes change smoothly into each other. These smooth figural changes can be inhibited somewhat in two ways. First, if the series of circles is alternated with the grid, the result is very often only flicker of two shapes, although replacement is also seen occasionally.

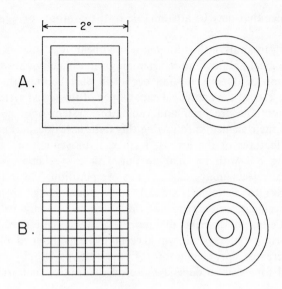

FIG. 7.5. Displays that yield plastic deformation or replacement, depending upon how they are positioned.

Second, when the arrays are laterally separated by one or two degrees, as illustrated in the figure, replacement rather than plastic deformation is reported. In these two cases, therefore, motion with replacement occurs as an alternative to plastic deformation.

Motion with replacement is not the only alternative to smooth plastic deformation in the plane. Still another perception, similar to what Orlansky (1940) first described and called figural fusion, is a composite of the two individual parts, that seems to move back and forth.[3]

Orlansky does not spell out the conditions for obtaining figural fusion, and some experimenters have not obtained it (Raskin, 1969). In our own experiments we have found occasional but clear instances of figural fusion when a Necker Cube was alternated with certain of its

---

[3] By "figural fusion" Orlansky means a composite of the two flashes; this is not the same as the fusion of two circles into a common one described in Chapter 4 as the opposite of fission.

parts (Fig. 7.4) and, again occasionally, when the laterally separated grids and circles were alternated (Fig. 7.5). The observer often sees a composite shape moving in the plane of the display. The finding speaks to a breakdown in the contour building operations the visual system engages in.

We can therefore distinguish at least three kinds of figural transformation in the plane: figural fusion, replacement, and plastic deformation. These distinctions are based on the appearance of the displays, hence are phenomenological categories. The likelihood is high, despite these differences in appearance, that they represent the action of a single mechanism responsive to small variations in the physical stimuli. Figural fusion occurs when the visual system attempts to construct representations of objects that are similar to each other on critical dimensions and thus require similar constructional routines. The timing is such that the routines used for the two constructions cannot be activated independently, or are not able to "shut down" completely before they are reactivated. The result is a visual fusion, a perceptual construction composed of parts of two things. Similar fusions occur in speech, especially when the speaker is tired or excited. Two intended words are expressed as a single "portmanteau" word, as "The man hates fish" becomes "The mates . . . ".

Motion with replacement occurs when the visual system is obliged to construct a shape that is markedly different from the first, using different constructive routines. Its occurrence indicates that the visual system cannot supply transitional figures that "rationally" connect the two termini. Hence, the visual system must regard such termini as sufficiently different categorially as not to be transformable into each other. Categorially different is not a matter of the names nor necessarily of topological difference. It speaks to the issue of "similarity", but is not yet well-defined. It is represented by the system's failure to achieve plastic deformation.

Plastic deformation occurs when the visual system is able to resolve the figural disparity between two flashes with a figure-building process that is already active. To put it another way, plastic deformation occurs when the figures are members of the same class in respect to figure-building operations but differ in detail; figural fusion occurs when the individual figures are members of the same class and are similar in certain details; and replacement occurs when the individual figures are not members of the same class.

Is there an empirical or theoretical method to establish what these class memberships are? It was with just this question in mind that Kolers and Pomerantz (1971) undertook their study of motion with disparate figures, but found that that method did not yield the answer. At the present time the only test that seems to provide evidence to the issue is the phenomenon of replacement, but for the reasons mentioned above—the brevity of the display and the difficulty of making the judgment—that phenomenon does not seem to provide the necessary opportunities.

The question may seem trifling and academic, but it is far more than that, for it speaks to the general problem of establishing a metric for visual shapes and to the topic of "similarity". For years psychologists have realized that we lack a rational way to describe visual figures other than by giving their measurements or their names. Measurements or even binary information about the location of dots in a matrix do constitute a notational system of sorts, but a tedious and ungainly one. A commercial television screen contains about 275,000 dots of phosphor that are excited 30 times per second. To describe even the shape, size, and location of a small polygon on that screen requires specifying a very large amount of information, far too large for efficient verbal use. Is there no method other than measurement, then, to describe even simple figures? One approach that has been taken, mentioned in Chapter 4, is to require subjects to judge the similarity of pairs of geometric shapes, or to decide which of two shapes a third shape is more like. The judgments are correlated with a large number of measurements made on the physical characteristics of the shapes themselves, and the correlations are factor analyzed. By these means, it is hoped, some suitably small set of characteristics will be identified that are important to judgments of figural similarity; these could be used to construct a metric and perhaps even a notational system for shape. Achieving these goals is a most desirable outcome, but success with this method has not been great.

One reason for the lack of success emerges from these experiments on apparent motion. Any shapes that the visual system can deform plastically into each other must be members of the same perceptual class in some sense, and we have found that the visual system can transform almost any simple line drawing into any other. Hence, the measurements that are made as part of the effort to establish a visual alphabet or notational system (a "visual grammar," so to speak) seem actually to be measurements made on members of the same set. To put it another way, the shape metricizers, in trying to define an alphabet, have been measuring only varied instances of the same letter, as a capital ay, lower case ay, roman ay, italic ay, cursive ay, and gothic ay are varied instances of the same letter; they have not been establishing the alphabet.

It is important to distinguish, therefore, between featural similarity and constructional similarity. A square and circle are markedly different in respect to their features, a fact that feature-analytic theories of perception and the shape metricizers make the unquestioned keystone of their argument. But if the visual system can transform a circle into a square as readily as it can move one circle to another, then the circle, square, and intervening stages must be instances of a single figure-constructing operation. The visual system uses, if not the same, then highly similar processes or routines to build the two shapes; it is engaging in constructionally similar operations to build what are featurally different configurations. Judgments of similarity and difference of features, therefore, cannot illuminate very much, if they can illuminate at all, the actual constructive processes the visual system uses; they probably cannot yield the basis of a visual alphabet. The features that we identify or judge in a configuration are the results of scrutiny and analysis, and are not necessarily the characteristics that the visual system uses in its construction of visual representations.

It is against this background that one may say that fusion occurs when the figures are members of the same class and share critical details; that plastic deformation occurs when the class membership is not also burdened with excessive detail; and that replacement occurs when the figures are not members of the same class.

The occurrence of replacement implies further that the visual system is sensitive to the configuration on its retina as a configuration, and not merely as a pattern of excitations. This assertion in the present context does not mean, as the Gestaltists meant, that contour is basic, fundamental, or given; rather it means that even though contour is

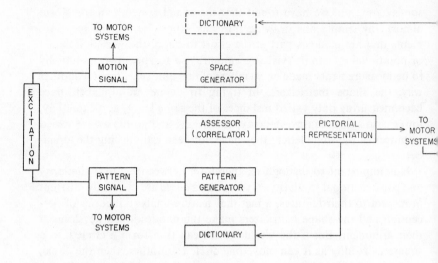

FIG. 7.6. Schematic representation of the visual construction of apparent motion. Dashed lines indicate uncertainty in the representation.

a relatively late component in the construction of visual stimuli, certain of its characteristics are monitored from the beginning nevertheless. In some way the visual system "knows" that a contour-building operation it is using with one figure will not be adequate to form another that is presented less than a third of a second later. The figural construction is not created from this knowledge, but the knowledge does affect the construction. Although the motions between figures can be accounted for as the obligatory resolution of a disparity between members, the disparity itself must be apprehended by the system and taken into account in constructing the resolution.

In Fig. 7.6 I have schematized the kinds of operations described. Excitation of the visual system generates two kinds of signals, H-signals (motion signals) and V-signals (pattern signals), some of whose characteristics were described in Chapter 5. The H-signal component of the V-signals is mapped by an Assessor, whose characteristics are described below. The V-signals are associated with such features as spatial density, grouping, intensity, and contour. The information transmitted, the result of feature-analytic operations, is largely instruction to a pattern-generating system that synthesizes the extracted information, smooths disparities, and supplements the physical givens out of its own store. In the "dictionary" appended to the pattern generator lie the means of classifying shapes constructionally. "Replacement" occurs when the dictionary does not have an entry that corresponds to the joint requirements of the two flashes. "Fusion" occurs when the refractory period of the generator exceeds the stimulus input. "Plastic deformation" is the normal operation of the pattern generator when the stimuli presented fall within the same constructional class, that is, require similar operations for the analog of their pictorial representation.

The Assessor, in mapping the locations of stimuli, also acts as a co-ordinator or temporal correlator. It is on the basis of information supplied by the Assessor that the pattern generator, having found no suitable entry in the dictionary, hurriedly creates "replacement".

The analog on the motion side of the diagram to a figure generator is a space generator. Whether motions are perceived in two or three dimensions depends upon the input to the space generator and the possibilities available in its own dictionary. Orientation of objects, their location in space, and their bi- or tridimensionality are functions controlled by this subsystem. The Assessor is conceived of as co-ordinating the operations of the figural and motion components of this system. As we shall see below, it may do more than co-ordinate.

Extending out from the Motion Signal and Pattern Signal analyzers are pathways to motor systems. The pathway from the Motion Signal analyzer is thought to control reflexive motions of the eye, the sharp and rapid orienting of the eye toward a flash or moving object detected by the periphery. As such motions occur on the basis of minimal amounts of visual information and usually are completed in the absence of clear perceptual representations of the object itself, it seems implausible that they require the elaborate processing that would occur if the signals travelled through to the Pictorial Representation. Similarly, on the Pattern Signal side, the elegant experiment of Fehrer and Raab (1962), and the subsequent control study (Raab and Fehrer, 1962) show that a motor reaction to a flash of light can be initiated long before the pictorial processing is completed. Fehrer and Raab attenuated the apparent brightness of a flash by masking it with another that followed the first in time. They found that the speed of reaction to the flash was related to its physical intensity, not its apparent brightness; subsequently, they showed that the reaction is programmed by sampling approximately the first 5 msec of the presentation. (The phenomenon of backward masking is described more fully in the next chapter.) This suggests again that certain aspects of the information in a signal can be processed by the ready observer in a manner that does not require that the information go through the channels that lead to Pictorial Representation.

Two questions of interest concern the means by which the spatial and pictorial dictionaries are stocked. We have very little evidence on this topic, so the following is offered entirely as conjecture. The extensive research carried out in the past decade on motion-dependent aspects of perception has indicated clearly that ambulation through the environment is required for certain kinds of perceptual adaptation. In particular the perception of depth seems to be inhibited severely when the organism is not allowed to ambulate freely in a lighted environment containing arrays of objects. One might conjecture therefore that the spatial dictionary is stocked by information fed back from the motor results of behavior carried out on the basis of the Pictorial Representation. That is, a sense of depth and other attributes of space are developed on the basis of the individual's activity and relation to objects that are perceived. The dictionary of pictorial options, on the other hand, could be stocked entirely on the basis of pictorial experiences alone. Hence, what one sees may be enough to permit one to see that object again; but what one does in respect to what one sees determines the contents of the spatial dictionary. Language affects these processes too, more the pattern component than the spatial one, probably, and some aspects of the influence of language on perception are discussed in Chapter 10. The means by which linguistic and pictorial information interact is so uncertain at the present time that such an influence is not even shown in the diagram.

The scheme proposed preserves the idea that motion and shape are mediated by different subsystems; it allocates orientation and space to a motion subsystem (thereby bonding firmly the relation of motion and depth); and it accommodates certain aspects of the difference between the processing of environmentally supplied and self-supplied information. Some of these topics will be picked up again in Chapter 12, but first we must make another detour to consider some other aspects of the perception of illusory motion.

CHAPTER 8

INDUCTIONS, I

ABSTRACT

In this and the next two chapters a number of inductions are described. Here we concentrate upon the influence other stimuli exert upon the illusory object, especially figural and "contextual" stimuli. Experiments purporting to show the influence of "phenomenal space" are reanalyzed and reinterpreted. A distinction having been drawn earlier between figure and motion, another is proposed between figure and space; it is suggested that the visual system creates a perception of space separately from the objects perceived to be filling it. Significant differences are illustrated between visual masking and visual apparent motion, and all of these data are used to support the general idea of stagewise processing of visual information.

In this chapter I shall describe several inductions of motion and depth. Inductions, in visual phenomena, are interactive processes in which some event P is perceived in one way when K is also present, and in another way when K is absent. Induction effects in vision have usually been measured as functions of the temporal, spatial, and luminance relations between the target stimulus and the inducing stimulus that affects its perception. Many more variables than these three affect the illusion of motion. Some are related to modes of fixation, but others are due to graphemic and semantic characteristics of the stimuli, to their spatial frame of reference, and to repetition.

To begin with the last, Wertheimer (1912) reported a marked effect of repetition. When two lines forming an angle were alternated many times at the appropriate rate good motion continued to be seen for a few trials even when one of the lines was suddenly eliminated from the display. DeSilva (1928) confirmed this result; presenting one of two stimuli that normally are paired creates conditions in which the visual system supplies itself with the information it has come to expect. Such self-generated perceptions are actually quite common both in visual perception (Shaffer and Wallach, 1966) and in other domains of behavior in which anticipation plays a role (Lashley, 1951). For example, if one says to an American, "The National Anthem of the United States is 'The Star-Spangled' ", the person can very often "hear" "Banner". Analogous effects occur in vision as partly perseverative, partly anticipatory constructions. Events of this kind are often taken as evidence for the constructive and anticipatory rather than purely receptive nature of perception: the perceiver is thought to generate or construct what he perceives and anticipates rather than only to detect signals. When he hallucinates, he does so because repetition, needs, expectations, or wishes induce him to construct from one or two clues an entire complex, but one that is not matched by subsequent signals from the environment. The action of needs and wishes, however powerful they may be, nevertheless has limits. No matter how hard they wish it, most observers cannot see certain forms of apparent motion; the stimuli must be appropriate to the construction, for the visual system is not infinitely plastic. Some further effects of repetition will be discussed in Chapter 10.

Spatial Context

Inductions in apparent motion occur at several stages of stimulus processing. One involves the spatial framework or context. A vivid example is illustrated by Array 44 of Fig. 8.1. Three circles in the first flash are followed by two in the second, one figure unit below. When the interstimulus and intercycle intervals are brief but equal, this display is similar to Array 18 of Fig. 5.3, and the motions described there can often be seen. On many other trials, however, a single row of three circles is seen to oscillate in parallel columns; but while the two flanking circles move to their mates, the middle circle seems to move only to an imaginary line connecting the tops of the lower pair of circles, from which point it moves in tandem with the other two back to the upper row. The effect is so strong that it can still be seen even when the centers of the three circles are about 2.3° apart, about the limit with the apparatus we used (Array 45). In Fig. 8.1 the extent of motion of the central circle is indicated by the dotted line. Although the spatial range of this inductive effect is quite large, the effect itself seems to be quite limited, as the following test shows.

In Array 46 the solid circles again represent the first flash and the hollow circles the second although, as heretofore, all of the circles in the experiment were hollow. The pair of circles on the right appears to move laterally, but they do not induce a similar lateral motion of the circle within the ring. The single circle within the ring of circles can be seen to move selectively in any of several directions, laterally, vertically, or diagonally, and sometimes it even disappears entirely, without motion, when it alternates with the ring of circles ("replacement"). Thus the horizontally moving pair of circles on the right cannot induce a direction of motion on the enclosed circle. Distance is not the explanation of this failure because distance between the inducing pair and the target circles in this case is less than it is in Array 45. Hence the induction effect in those arrays is not due to inability to perform the proper spatial analysis, but its effective cause is not known. Whether direction of motion is important, vertical in Array 44 and 45 but horizontal in Array 46, is also not known. In other cases the induction effect is even more subtle, and is due not to the effect of one object upon another, but to the perception of an object as lying within a spatial matrix or frame of reference.

Perhaps the best-known work along these lines are the experiments of Duncker (1929; Wallach, 1959) on what he explicitly called induced motion. [DeSilva (1926) had a similar idea when he studied what he called "encompassment".] In the paradigmatic case, a small target is displaced relative to a larger framework; irrespective of whether the target or the framework actually moves, it is the enclosed target that always seems to move. The frame provides a spatial reference or context within which the motion is seen. Under most circumstances, when there is a displacement between the two, the observer's visual system assumes the framework is steady and the target is moving. Duncker argued therefore that perceived motion is always relative, that there is no absolute stimulus for motion.

The argument can be extended. Lacking other information, the visual system sometimes even takes itself as its frame of reference. In a totally dark room, a small point source of light that is fixated steadily soon is seen to move. Apparently, the eye does not record its own small motions; excitation by the stationary point of light of different retinal loci during those small motions is interpreted as the eye the framework and the point source the target. Although the evidence is not yet perfect

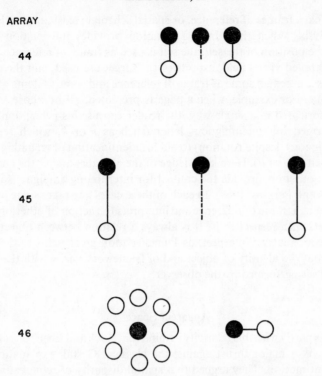

FIG. 8.1. Arrays 44–46. Successful and unsuccessful inductions.

along these lines (Royce, Carran, Aftanas, Lehman, and Blumenthal, 1966), good support for this analysis comes from the experiment by Matin and MacKinnon (1964). They found that when the relative motion between eye and point source in one direction of motion was eliminated by optical compensation, motions in that direction were no longer seen, but motions continued to be seen in other, uncompensated directions.[1]

[1] A possible limitation on the eye-movement explanation of autokinesis is the occurrence of depth effects. I have found that very long observation of a stationary point source in the dark (about 45 minutes or more) yields perceptions of the point source as receding or approaching—that is, moving in the third dimension. This occurred even when the observations were made monocularly, thus ruling out the effects of convergence.

Contexts, frames of reference, or spatial schemata exist at many levels of analysis. When the stimulus situation provides information only about the position of the self, the self is used as frame of reference. In a more detailed visual field, aspects of the target are used, and the components are treated some as frame of reference and some as items within the frame. For example, when a page is presented, all of whose letters are misoriented in a single way, the reader can use his perception of a transformed but unambiguous letter (such as *r* or *k*, which remain unambiguous despite rotation) to aid his identification of an ambiguous one (such as *p* or *d*). Here knowledge of the alphabet and of the characteristics of letters provide the context for interpreting ambiguous items (Kolers and Perkins, 1969). In each of these cases one takes some object or event as a frame of reference and interprets the action of other objects or events with respect to it. It is always a relation between object and framework that we interpret, as Duncker took great pains to demonstrate, but the identity of object and of framework varies with the level of analysis performed by the observer.

Apparent Space

It was partly with this relativity in mind that Rock and Ebenholtz (1962) undertook a major thrust against the classical Gestalt explanation of apparent motion. They argued that spatial disparity of retinal stimulation was not a necessary condition of apparent motion; sufficient in many cases, as the preceding chapters show, but not a necessary condition. The classical theory asserts that stimulating disparate retinal regions produces radiations in corresponding cortical regions; a short-circuit between the cortical regions is the neural correlate of the perceived motion. Many writers disputed this theory, with demonstrations of depth effects (Neuhaus, 1930) and of good apparent motion with flashes presented to heteronymous portions of the retina, and thus to separate cerebral hemispheres (Smith, 1948). Rock and Ebenholtz went farther. They argued that a perceived change of framework with no change of retinal position of the target flash should induce a perception of motion in the target, and reported finding this result.

Their main experiment required the observer to move his eyes back and forth across a dark surface in synchrony with the flashing of lights at two neighboring apertures. If perfect synchronization were attained, the two flashes would each stimulate the same portion of the retina. According to the classic Gestalt theory no corresponding neural interaction would occur, but their observers reported good apparent motion nevertheless. Rock and Ebenholtz then argue that it is change of phenomenal location (change in perceived space) not change in retinal location that accounts for apparent motion. Their arguments are interesting, but their experiment does not justify their conclusions, as we will see.

Several recent experiments have investigated various aspects of the perception of location and found it to be a complex function of time, space, and assumptions (instructions). The functions are complex enough as to make untenable any notion that perceptual space stands in a simple relation to the cartesian matrix we utilize descriptively (Matin, Matin and Pearce, 1969; Matin, Matin and Pola, 1970; Kinchla and Allan, 1969). The alternative, however, to attribute motions to "phenomenal space", noting its difference from cartesian space, is not a sufficient answer, although it does point to a line of analysis that needs following. First, the way frameworks operate to affect the perception of targets within them is not clear, as the experiments illustrated in Figure 8.1 amply demonstrate; and second, motor systems, and especially movement of the eyes, strongly affect perceptions.

A rich lode of phenomena has been exploited first by the sensory-tonic theorists (Allport, 1955) and more recently by investigators who require subjects to learn new perceptuomotor co-ordinations when the visual input has been optically transformed (Harris, 1965; Howard and Templeton, 1966; Freedman, 1968). In an example immediately relevant to the present context, subjects were instructed to move their eyes clockwise or counterclockwise while a directionally ambiguous display, crosses that seemed to rotate in either direction equally, was flashed before them in apparent motion (Pomerantz, 1970). The observers' task was to indicate the direction in which the crosses seemed to move. The finding was that more than twice as often they seemed to move in the same direction the observer was moving his eyes as in the opposite direction. When fixation was maintained on the center of the figure, motion seemed to go clockwise or counterclockwise about equally.

Hence, eye motions do not completely determine the direction of perceived motion (although it is not known whether, despite efforts at fixation, the observer's eyes moved involuntarily nevertheless), but they do affect the interpretation of perceived motions rather powerfully. This fact, interesting in its own right, restricts the interpretation of the Rock and Ebenholtz experiment.

A related effect restricts the interpretation of an induced motion studied by Brosgole (1966). He arranged two arrows, illustrated in Fig. 8.2, so that the small line common to them was superposed. He reports that when observers see the alternated arrows in three-dimensional motion, the apparent location of the center dot changes. Brosgole attributes the apparent motion of the dot to the change in context, or

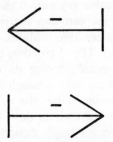

FIG. 8.2. When the shafts of the arrows are superposed and the two arrows alternated, the apparent location of the small marks above the shafts changes under certain conditions of fixation. From Brosgole (1966).

change in phenomenal distance of the dot from the two ends of the line. In repeating the experiment we have found, however, that fixation locus is very important. When the observer fixates either arrowhead or either line, a clear and compelling lateral motion of the repeated dot is seen, confirming Brosgole's observation. But if the observer fixates the dot itself, no equivalent motion is seen. Thus, in addition to its dependence upon locus of fixation, this induced motion effect seems to be similar to that reported with Arrays 44 and 45: in this case the plastically deforming heads of the two arrows induce an apparent motion in an intermediate shape. It may not be necessary to appeal to changes in phenomenal space to understand either the Rock and Ebenholtz or the Brosgole experiment. Motor influences and spatial inductions at the graphemic level may do the job.

Inductions are not restricted to motion in the plane. In Chapter 6 it was shown that a pair of squares that normally moves in the plane of the display can be induced to move in depth when they are presented simultaneously with trapezoids that are themselves seen in depth. That demonstration revealed that the induced motion can affect not only the interactions of the contours themselves, as in Arrays 44 and 45, but also the mechanisms that mediate the perception of contours in depth; that is, a spatial map.

Spatial Co-ordinates

Some experiments suggest that the visual system represents all flashes upon a simple two-dimensional matrix, and carries out very little if any stimulus analysis of the objects upon that matrix (Fig. 8.1). Other experiments suggest that the spatial matrix is at least three-dimensional, and that the visual system does perform some analysis of the objects within it, as revealed for example by the perception of a square and trapezoid (Fig. 6.3), moving one in the plane and the other in depth. The way in which the visual system represents the spatial aspects of objects, and differentially creates the objects and their space, are continuing puzzles. Several experiments suggest that the moving stimulus and the space in which it is represented are constructed as different aspects of a perception, much as feature-analysis and pattern synthesis are different aspects (Fig. 7.6). Let us consider two.

If one stares at an arithmetic spiral (Archimedes spiral) while it rotates at three revolutions per second, compelling illusory motions are seen, sometimes even in depth. If one stares for a minute or more and then immediately transfers his gaze to some other object, or even continues to stare at the spiral after its motion has stopped, a powerful aftereffect is seen. If during its rotation the spiral had seemed to expand, the objects looked at subsequently seem to contract, and if the spiral had seemed to contract, objects looked at subsequently seem to expand. Wertheimer (1912) thought that the illusory motion obtained with flashing lights and the illusory motion obtained with the spiral were equivalent instances of illusory motion; and even more recently Sekuler and Ganz (1963) have regarded the spiral in approximately the same way. This analysis is faulty, however.

The fact that one obtains a powerful effect after observing a moving arithmetic spiral, or a waterfall, cannot by itself be used to prove that illusory motion creates the effect. The reason is that movement of the contour is itself quite real; the physical position of lines changes with time. It is the sense of contraction, expansion, or depth that is illusory but not the movements. If one eliminates the contour and merely exposes two or three stationary points at the same temporal period they would have in a rotating spiral, obtaining thereby good optimal motion, no aftereffect of motion is seen, as I have found in many tests and as Humphrey and Springbett (1946) also report. In my own experiments observations were made with the interstimulus and intercycle intervals equal, and with the intercycle interval between 2 and 20 times longer than the interstimulus interval. Observation was maintained for as long as 2 minutes, with the eye fixated or with the eye free to wander. Immediately afterward a stationary target was presented, but no distortion of its appearance was reported. Therefore, although alternated flashes yield a compelling perception of motion, they do not create a motion aftereffect. A contour moving physically across the retina seems to be required for the aftereffect.

The effect of the moving contour in the spiral illusion is sometimes quite extraordinary. In another test (Kolers, 1966) observers first noted that a large matrix of squares was perfectly regular. The matrix subtended about 20° of visual angle on each side, and the cells within the matrix subtended about 0.5° each. After a minute's inspection of a rotating spiral that subtended about 4°, observers transferred their gaze to a single cell, approximately in the center of the matrix. If the spiral had rotated so that it seemed to contract, the cells of the matrix that now fell within the 4° region that the spiral had occupied seemed to bulge, in two and sometimes three dimensions. However the remaining lines of the matrix appeared straight. Moreover, if fixation was maintained carefully, no discontinuity was perceived between the enlarged cells and the remainder of the matrix, a result which, from the geometric point of view, is, of course, an impossibility.

If the moving spiral had affected some perceptual center concerned only with the representation of contours, then the central cells of the matrix would have appeared distorted, and disconnected from the remainder. If the moving spiral had affected some center concerned only with the representation of motion, then the cells might have pulsed in and out or rotated in the plane of the display. Neither of these possibilities was observed. Therefore a different sort of explanation is required.

It was shown in Chapter 5 that a contour that lies close to or in the path of other contours that are seen in motion will induce a perception of depth in the apparently moving figure. The idea that depth follows or precedes motion, or that depth comes from motion was rejected, and in its place was offered the alternative that depth is the perceptual result when neighboring regions are excited within proper temporal bounds. (By analogy, hydrogen and oxygen do not precede water in any logical way; water is hydrogen and oxygen in proper combination. Decomposition and precedence are not logically equivalent notions.) When contours properly placed move in the neighborhood of others, the interactive result is signaled in the visual system as the analog of depth. My proposal is that the moving spiral does its work not on the motion system itself, nor on the contour system itself, but on the interaction of figural processes and the space generator of Fig. 7.6. In two-dimensional apparent motion the H-signals dominate and constrain figural interpretations. In the spiral aftereffect, contours themselves, and especially such of their characteristics as proximity of parts and disparity of orientation, induce the opposite result: aspects of figure constrain the spatial representation.

The implication of this analysis is that space is a perceptual event that is constructed somewhat separately from the objects that are perceived to fill it; hence some apparent spatial displacements of objects may actually be due to a deformation of the space in which the objects are perceived to lie rather than to a deformation of the objects themselves. Many of the phenomena described as *figural* aftereffects by Köhler and Wallach (1944), for example, seem to be better described as distortions of the space within which figures are perceived rather than as distortions of figure.

Wertheimer's use of a rotating spiral to demonstrate the aftereffect of motion, therefore, is probably not justified, for there not motion but figure seems to be the source of the aftereffect. Although Russell (1945, p. 806) argues that "motion is made out of what is moving, but not out of motions", his distinction is logical, not neurological. For the visual system motion is indeed sometimes made out of motions, as in figureless phi, and sometimes out of the object moving; but the motion itself, unlike the object moving, seems to leave no after-discharge.[1] The rotating spiral inseparably confounds motion signals and figure signals, and it is the latter rather than the former that produce the aftereffect. The perception of objects and the perception of their motion are operations undertaken separately in the visual system and are experimentally dissociable. So too are motions and the space in which they seem to occur. The perception of an object moving is a phenomenological reality; its unity is achieved, however, not as the given the Gestaltists invoked, but as the consequence of an integration of several subprocesses.

Having seen that phenomenal space need not be implicated to understand induced motions, let us examine another cognitive constituent, the semantic level of analysis. The question is whether semantic features of the stimulus affect the perceptibility of motion.

Semantic Factors

As shown in Chapter 4, well-learned and well-formed identity, that is, the semantic level of figural analysis, can modulate somewhat the apparent motion effect. Of course, shapes readily lose their identity when they are presented in properly timed sequence with other shapes, and do so the more as their durations are increased. In those cases, however, we were dealing with plane figures, and it is possible that all plane figures are members of a single class of object, hence can interchange their identities readily. But what if a plane figure is alternated with letters, say, or with words? Letters and words are plane figures or pictures of a kind, but they are also symbols, for those who know them. Do their properties as symbols affect apparent motion?

If a difference in identity can induce the visual system to inhibit the perception of motion, the effect should be most marked when the stimuli are members of different analytic classes of objects. To test the possibility we alternated shapes such as a circle and the letter *a*, or the word BOX enclosed in a circle or rectangle, or the letter *x* with the letter

[1] Except perhaps as in the reverse motion described in Chapter 5.

a. In none of these cases was motion inhibited. An *x* does not change as smoothly into an *a* as will *e*, but motion with either plastic deformation or replacement is always seen. Similarly, as reported in Chapter 4, a box surrounding BOX or CAT moved smoothly and continuously to the subjective border of the word, where then some plastic deformation and some replacement occurred. Despite the disparity of level of analysis of the stimuli—shapes as plane figures compared with shapes as words —motion between the shapes always occurred, and was dominated by the graphemic features. The word BOX and a rectangle share more semantic features than the word CAT and a rectangle do, yet the same perceptual results were obtained in both cases. Hence, semantic features or "meaningfulness" seem to play little if any role as an inhibitor of apparent motion in these tests. Semantic features may play some role as a facilitator, as we shall see later. In general, however, the visual system seems to perform little if any analysis of the stimuli at the semantic level before creating the perception of movement. The excitatory pattern created by duration of the flashes and distance between them, but not their semantic relations or semantic disparity, are the primary determinants of the motion effect.

When meaningfulness plays a role, its chief effect is on the clarity and vividness of the motion perception, upon its interpretation, but not upon the occurrence of the perception itself. Just as expectations can induce the practiced observer to see motion when only one figure is presented (Wertheimer, 1912; DeSilva, 1928), knowledge of and familiarity with the stimulus array enables the observer, in a manner not yet understood, to make a more efficient, more rapid, or more comprehensive representation of the motion than he otherwise would. It is in this sense that the effect of meaningful stimuli upon the perception of motion (DeSilva, 1926; Steinig, 1929; Toch and Ittelson, 1956) is to be understood. To put the matter another way, measuring the probability of seeing motion as a function of duration and interstimulus interval should not be the method of choice for assessing the influence of semantic factors upon the perception. Those parameters are best able to reveal the interaction of H-signals, but the influence of semantic factors would best be made clear by an analysis of the V-signals. "Meaningfulness" is a matter of interpreting patterns, not of extending temporal boundaries. Perhaps clarity, convincingness, or other aspects of recognition, as studied in earlier days (DeSilva, 1928) would better reveal the action of semantic features than timing would.

Knowledge of what the stimuli are composed has little effect upon the temporal characteristics of apparent motion. Skilled observers who know, even, that the illusory motion is created by the flashing of lights cannot with that information alone inhibit their own seeing of motion. Recently Segal and Barr (1969) confirmed that whether subjects were told they were seeing an illusion or not made little difference to their measured threshold for distinguishing motion from simultaneity.

We may say therefore that the major inductions found in the study of illusory motion are the results of patterns of excitation. Knowledge, familiarity, and meaningfulness of the stimuli play, if any, only a very small role in varying the temporal characteristics of the phenomenon; but it remains an open question whether and if so how they affect more cognitive components of the illusion, such as the clarity, convincingness, or recognizability of the objects in motion.[2]

Observers are of course sensitive to the fact that BOX spells a word; therefore, the fact that that word is seen to go into motion with a line or circle suggests that in some sense the perception of BOX as a word is accomplished by different and perhaps later routines than those responding to BOX as a shape. Many contemporary workers believe that the visual system does its work by processing information in stages; indeed, the whole argument concerning whether depth precedes or follows motion (Julesz, 1971) has just this sort of model in mind. The issue under discussion concerns the way in which the unity of visual experience is actually comprised of the action of a number of subprocesses or stages. Experiments to be described make more credible the finding that some aspects of a configuration may affect the motion perception while others do not.

---

[2] Brenner (1953) argued against the Gestalt theory by showing that people can do mental arithmetic at the same time that they can see apparent motion. We can only conclude from this study that performing mental arithmetic and seeing apparent motion are activities carried out by different parts of the brain. One can also eat while doing mental arithmetic and whistle while reading (Huey 1968). Only the most naive view of the brain's organization would find Brenner's results surprising.

Masking and Motion

Over temporal intervals and spatial arrangements that are similar to those characterizing good beta motion, a curious brightness and contour-inhibiting process also occurs. It has come to be called backward masking when the observations are made on shapes, but a variant of it had earlier been studied by Dodge (1907), who spoke of "clearing-up time", and other variants related to inhibition of brightness are called metacontrast (Alpern, 1952) and the Crawford effect (Crawford, 1947; Boynton, 1958). In all of these cases, but in different ways, the perception of the first of two brief visual presentations is interfered with by the presence of a second. In a classically simple demonstration of backward masking for figures, Werner (1935) presented a small black disk to the eye briefly and followed its offset with a black ring. Had the shapes been presented simultaneously, the ring would have just circumscribed the disk. When each shape was presented alone, each was clearly visible. Werner found that at certain temporal separations between the flashes, the presence of the disk went unreported, the disk having been "masked" from perceptibility by the aftercoming ring. This trivial but theory-challenging phenomenon has been studied intensively in the past dozen years (Raab, 1963; Kolers, 1968; and Kahneman, 1968, have reviewed various aspects of the data), and several workers have noted similarities between the temporal intervals that yield masking and the intervals that yield good apparent motion (Toch, 1956; Fehrer and Smith, 1962; Kahneman, 1967). Some workers have suggested that they are related aspects of the same phenomenon and Kahneman (1967) has alleged that masking and metacontrast are the result of "failed motion".

Several facts indicate that attributing one of these effects to the other is impossible on the face of it. Consider the general situation. The physical environment is, informationally, infinitely dense; perceptual experience only samples the density at any time. Hence, perception is typically an information-losing or even, considering what can be on the retina, an information-destroying operation, for less is represented pictorially in awareness at any moment than is available from the optical projection on the retina. Backward masking is an example of an information-destroying operation, for when it is masked, a normally visible object is lost sight of. Apparent motion, on the other hand, is an information-creating operation, for more is then represented in perceptual experience than is projected on the retina. This operational difference between the two phenomena is accompanied by other differences. I shall list five.

First, it has been pointed out (Kolers, 1968) that varying stimulus duration has different effects on the two phenomena: the longer the duration of the first flash in a masking sequence, the less likely it is to be masked, but the longer the duration of the first flash in a motion sequence, the more likely it is to be seen in motion. Second, increasing the duration of the second flash increases its masking of the first, but duration of the second flash plays almost no role in apparent motion. Third, the masking effect falls off sharply as distance between the two flashes is increased. The maximum distance between the borders of the flashes at which masking has been reported is less than 2° of visual angle (Alpern, 1953; Kolers and Rosner, 1960); good apparent motion is seen at spatial separations more than twice as great. Fourth, the masking effect is markedly sensitive to similarity of features in the masked and masking objects; the more nearly alike and the closer in space the two contours, the better the masking effect (Werner, 1935; Schiller, 1966). The wide range of disparities of shape that plastic deformation in apparent motion can resolve reveal no such restriction. And fifth, Kolers (1963) reported that a black disk masked by an aftercoming ring can still be seen in motion with another disk. The first disk may be made partially or wholly invisible by the ring but good motion to the other disk is still reported. The simultaneous occurrence of masking and motion makes it impossible that masking is failed motion.

A wholly satisfactory theory is still not available either for masking or for apparent motion. However, differences in the way the two phenomena are affected by variations in stimulus conditions make it clear that one is not a form of the other. Both phenomena can reveal transformations or alterations of contour, but masking seems to be a genuine contour-interaction effect, a phenomenon almost exclusively of the pattern signals and pattern generator of Fig. 7.6. Apparent motion cannot be described that way. The interval of about one-third of a second bounds both motion and masking. It bounds a number of other perceptual and productive processes, both pictorial (Kolers, 1963) and linguistic (Broadbent, 1958; Kolers, 1966). In the visual system we find that multiple flashes whose duration is less than that interval create conditions that, historically, the system seems never to have found it necessary to resolve accurately. The consequences are forms of illusion such as temporal integrating functions, backward and forward masking, and apparent motion. All of these phenomena reveal different aspects of the construction of perceptual events so that analyzing them sheds light on what the system normally does do. One hopes that at some time all of these phenomena will be accounted for by a superior theory of visual pattern recognition, but calling them by each other's names is not that theory.

It is worth noting, however, that there is some ambiguity of usage in the term "masking" or "backward masking". I used it above to refer to a specific interaction of contours, but the term can also be used (and has been used) to refer to any interference with the perception of the first of two displays. Partly as a consequence of the ambiguity, much heat has been generated in the recent literature (especially in the *Journal of Experimental Psychology, Perception and Psychophysics*, and *Psychonomic Science*) regarding the true shape of the function for backward masking. On close inspection it turns out that many of these articles describe experiments that are fundamentally different in kind from others (often not so noted); the result is that many comparisons are made improperly.

In different experiments the subject has been required to name the shape that was presented (such as a letter), to report on the integrity of its contour, to report on variations in its brightness, or to report, using forced-choice procedures, on its spatial location or temporal location within a sequence. These are not only different experimental procedures, they require markedly different judgments. The different judgments are probably based on the processing of different aspects of the stimulus. Hence there is little basis for generalizing from one experiment to another.

Under certain circumstances the curve for brightness matching and for contour is a U-shaped function of the time between the target and masking flashes; but in other cases the interference is found to vary monotonically with time. The ratio of luminances between the target and masking flashes (or, for black forms, the corresponding contrast relations) are found to determine whether the U-shaped or monotonic relation is found (Kolers, 1962). I showed, moreover, how those ratios can be manipulated to transform a U-shaped function into a monotonic one, and vice-versa. A similar analysis has not, to my knowledge, been carried out for masking of names or locations. The evidence we have seen in earlier chapters indicates, however, that information about location has a lower input threshold than information about contour or brightness, in the sense that one can tell where something is on the basis of less input than one can tell what something is; and one can tell what its shape is before one can make correct statements about its brightness or color. As a consequence of these differences, the interference exerted by an aftercoming flash depends upon the features the subject is instructed to report and upon the composition of the masking flash.

Some investigators have been remarkably insensitive to these relations in the stimuli and to the requirements of the task, and have made sweeping pronouncements (sometimes accompanied by vituperation) about what masking is. Obviously it is not "masking" that is or is not something of interest (except as a convenient term); rather it is the nature of the interaction between features of the target and of the aftercoming display, and the interaction between the displays and the subjects' judgments that are of interest.

The sweeping and frequently repeated assertions about what masking is create an embarrassing situation for us. Because they are repeated often, many people accept them; yet if we were all to accept them, we would be reduced to having a science based on fiat and repetition rather than a science based on analysis and evidence.

The phenomenon of masking does, however, raise an interesting theoretical question in respect to the diagram of Fig. 7.6. The question is whether still another feature, motion itself, can be masked. There are some reports in the clinical literature that describe distortions of motion perception, fine cerebral lesions that purportedly create truly Zenonian perceptions. Teuber (1960) and Teuber, Battersby, and Bender (1960) remark on patients whose perception is more like a rapid sequence of still photographs than like smoothly continuous motion; the figural component of their perception remains unaffected apparently while the motion component is lost. (These occasional reports, if validated, would of course be strong evidence for the separability of the two systems.)

One must ask therefore whether the perception of motion can be masked or distorted by psychophysical means alone, in the intact observer, as the perception of figure can be. Pomerantz and I made some preliminary experiments along this line, but it was not our good fortune to find a way to achieve this end. We found that H-signals will not cross paths, and we were able to reduce the lateral motion of a string of shapes to local plastic deformations (Fig. 4.5, Array 6), but we were not able to stop the perception of motion altogether by surrounding the target or by interposing other objects into the path of motion, methods that inhibit the V-signals in backward masking of shapes. Finding a psychophysical means to block the perception of motion without affecting the perception of figure would greatly facilitate the study of these two components of perception, and of their interaction.

Proof of Stages

Another experiment reveals further the dissociability of operations and the clear distinction that must be drawn between phenomenological experience and operational construction. Visual masking, as just described, is an induction extended in time. In the static case the visibility or apparent brightness of one contour is studied as a function of characteristics of neighboring contours. When apparent brightness is measured, the usual result is that the more intense the inducing contours that flank a target, the dimmer the target seems. In experiments on masking and metacontrast, the target and inducing contours are presented at different times, and the induction is studied as a function of the temporal separation between the flashes. When the inducing flash is a large disk of light on which the target is superimposed, maximum inhibition of the target occurs when the two flashes are presented simultaneously (Crawford, 1947). The analogous experiment can be performed dynamically, the target stationary and the inducing object a moving line. If the target is a small line of light and the inducing flash a larger line whose path moves it across the target, threshold brightness of the target can be found when the moving line is close or distant. If the target is presented for only a few milliseconds, the moving line of light inhibits the brightness of the target line more when they are close together in time and space, and less when they are separated (Luria and Kolers, 1962;

Luria, 1965). This is a straightforward induction, approximately similar in its result to masking with stationary flashes.

Suppose however that motion of the inducing line is illusory, created by the flashing of two lines of light that flank a small target. Under suitable circumstances optimal motion looks like or is phenomenologically equivalent to motion of a real line. An experimental question then asks whether the similarity of appearance of real motion and apparent motion extends to a similarity in the way the visually moving objects affect the target in its path.

The question was answered (Kolers, 1963) by flashing a small target light briefly in the path of a line in apparent motion, at an intensity that enabled the observer to detect its presence about 90 per cent of the time when it appeared alone. The two lines whose flashing created an illusion of motion were presented for 50 msec each, and the interstimulus interval was fixed at 105 msec. The target line was presented for 5 msec at different times during the interstimulus interval, that is, when the illusory line had seemed to traverse various portions of its path. The two conditions of motion, real and illusory, were perceptually similar, but the results were different. Unlike the control condition with real movement, the illusory movement made no difference whatever to perceptibility of the target. The similarity of appearance of real and apparent motion notwithstanding, the phenomenologically equivalent moving line affected the brightness of objects in its path differently in the two tests.

A related aspect of these results raises the broader question whether one spatial illusion necessarily partakes of the same processing as another. A well-known illusion is the apparent reversal of perspective of a Necker Cube that occurs with its extended observation. Another is the spatial translation of apparent motion. Both of these involve spatial displacements of a phenomenological object. If both kinds of displacement were due to a common mechanism, the two illusions should interact so that, say, a Necker Cube changed its perspective more when it was seen in apparent motion than when it was stationary. Just this outcome was obtained when two cubes were seen in apparent motion (Kolers, 1964). But when a single Necker Cube was only flickered on and off at the same rate that yielded apparent motion when it was paired with another, the single cube also changed its apparent perspective more than a stationary cube, and about as frequently as when it was seen in motion. Thus the outcome was not due to an interaction of perceived illusions; rather, a rate of intermittency that induces one illusion, apparent motion, also maximizes the occurrence of another, change of perceived orientation. Intermittency of stimulation, not interaction of perceived results, yielded the outcome. The perceptions were the consequences of the interaction but not their cause. The interactions of all three kinds just described occur at earlier stages of processing than conscious representation. Explanation of the means by which our perceptions feed back to affect our behavior should not be confused with explanation of the construction of our perceptions themselves. Pictorial representations are the output of the visual system; they are perceptual products, not the elements of perceptual construction.

If a person sees something and then does something about what he has seen, it is obvious that his behavior can be described as a sequence of events, in this case, perception and then action. In addition, we can demonstrate that the construction of a perception itself follows a sequence of events, or stages of processing. By "stages of processing" we mean that as the consequences of stimulation pass through the nervous system, from place to place, each place "extracts" a particular kind of "information" for operation. Stages in this sense implies a system that is serial with respect to locations in the nervous network. For example, from physiology we know that information operated upon by the retina is further operated upon by the lateral geniculate bodies, and that in turn is operated upon by the primary projection areas of the brain, then the peristriate areas, and so on. A system that is serial with respect to location, however, can also be parallel with respect to function. Again from physiology, we know that the lateral geniculate sends fibers not only to the cortex but also to the cerebrum, and the latter in turn sends fibers to the cortex (Fadiga and Pupilli, 1964).

The idea of stages of visual operation has been available in psychophysical research for a very long time. Binocular processing of aspects of contour has been empirically known to occur at least since Wheatstone's (1838) days; but, as Sherrington (1897, 1906) showed, luminances do not interact in the same way. Hence, the processing of luminance seems to be carried out principally in the visual channel stimulated, but the processing of contour is less restricted in this way. Analogous results on visual masking have been documented (reviewed by Kahneman, 1968; Kolers, 1968): masking by a flash of light is almost entirely restricted to the eye stimulated, but masking by a pattern or contour can be demonstrated when one shape is presented to one eye and the second to the other. Color vision too seems to depend upon processing that occurs very early in the visual channel, and, like masking by flashes of light, reveals, if any, only trivial amounts of interaction of signals presented to the separate eyes.

Although available in the psychophysical literature for a long time, the concept of stagewise operation has received its strongest thrust from electrophysiological experiments. It is known at the present time that the visual system contains cells that are selectively responsive to properties of the stimulux flux, to its wavelength, intensity, motion, and the like. These cells are thought to be arranged in a hierarchical manner so that each node of the structural tree operates on special attributes of the input. For example, at lower levels of the network cells respond unselectively to the intensity of the stimulus, but at deeper or higher reaches of the network cells may respond selectively to movement, and at a higher level still only to movement in a particular direction. Hence, as one advances into the hierarchy, cells are found whose response is increasingly selective, refined, and specific, so that at each stage of the network some "editing" of the input occurs. Information is extracted "digestively", as it were, as the consequences of the optical image pass through the many places of the neural system.

The idea of stages has importance for our understanding of perception not only at the electrophysiological level, but at the psychological level as well. We distinguish stages in the construction of a visual perception, but we must also distinguish states of the percept. Stages refers to the various operations the visual system engages in when transforming physical stimulation into a perceptual outcome; and both electrophysiological and psychophysical evidence speaks to this issue. By states I mean to distinguish between the construction, maintenance, and decay of a perceptual experience. Most of this monograph is concerned with construction, and stages of operation within a construction. We know very little about the processes characterizing maintenance, although this may be the chief operation of visual perception; consequently only a brief discussion of this topic is possible, and that is postponed to the Afterword. As to decay, that is a topic of its own, involving aftereffects of prolonged observation usually, but a word concerning one aspect of decay follows here.

If by suitable means an optical image is made to remain stationary on the retina, the perception changes rapidly with duration of view. At first the whole object is seen, but then its parts fade in a more or less orderly way. At one time it was hoped that information about the way objects faded from view would reveal how the visual system constructs perceptions (Pritchard, Heron, and Hebb, 1960). This hope too has faded, however, for the rules governing decay under optical stabilization are not necessarily the rules governing perceptual construction. Only in very simple mechanical structures is the syntax of disassembly likely to mirror the syntax of assembly; in dynamic structures involving transformations, it is less likely to. Therefore, the facts obtained from studies of perceptual decomposition—facts regarding the importance of curves, planes, edges, and the like—are not thereby guaranteed to be facts important to figural constructions. In the one case a sequence of operations occurs; in the other, something fails. Neither the elements nor the procedures are likely to be the same.

We come then to the illusion of motion created by flashing lights. The perception obtained "looks like" real motion. The flashing lights generate information in the visual system that in some sense satisfies the requirements of a certain stage of visual construction; but at another stage of operation the forgery is recognized: interactions such as masking that can occur when a line of light is in real motion do not occur when the motion is illusory. The requirements at an early stage are less precise than those at later stages. Although not much can be said about the details of formation of the motion experience from the experiments just described, we may infer from them that the visual system is itself arranged to operate on different aspects of the visual signal in some kind of order.

We may infer too that the "same" phenomenological reality—as a perception of brightness, contour, or motion—can be constructed by different means. Hence the experiments reveal that as with depth, so too with motion: the "same" phenomenological reality can be created by considerably different means. Analytical experiments of the kind described can sometime illuminate the differences in workmanship and strategies of construction. Inductions, as they are often illusions themselves, can aid in the dissection and analysis of the operations or sequence of stages. Some other kinds of induction than luminance and contour are discussed in the following two chapters, and the concept of stages is discussed further in Chapter 12.

CHAPTER 9

INDUCTIONS, II

ABSTRACT

Under certain conditions of timing and placement, the illusory object itself undergoes radical alterations in appearance. Other alterations are associated with the modality stimulated. The main point made is that the object perceived in apparent motion is an accommodating rationalization: figural processes are highly plastic.

The inductions discussed in the preceding chapter reveal the effect of a second stimulus upon the perception of the target in motion, and the effect of an object in motion upon the perception of the target. In the present chapter the inductions discussed reveal the effect of motion upon the perception of the target itself.

One form the interaction takes affects the apparent location, size, and shape of the test object. Two lines of work have been reported, one with illusory motion and the second with physically moving stimuli.

Spatial Distortion

The phenomenological basis of some investigations of illusory motion is that the apparent distance apart of lights whose flashing yields the perception of motion seems to vary with the interstimulus interval; this finding has suggested to some investigators that spatio-temporal relativity characterizes some operations in the visual system. In one case, the halves of a bisected physical distance ABC appear longer or shorter depending upon the temporal relations of AB and BC. Scholz (1924) reviewed the earlier work and undertook a quantitative examination of the effect and its relation to the temporal intervals associated with apparent motion.

Scholz simplified the situation. He arranged his apparatus so that

141

two flashing lights yielding simultaneity, optimal motion, or succession, appeared a little above two continuous lights. On successive trials the distance between the flashing lights was varied until the subject reported that it matched the distance between the continuous lights, which was itself varied in several experiments from about 0.3° to about 17.3°. Scholz's major finding is that at small angular separations an expansion of their apparent distance is induced by the flashing, but at large separations a substantial contraction occurs. The magnitude of both effects varied with the interstimulus interval between the flashes, and was maximal at the same interstimulus interval that yielded optimal apparent motion. Scholz also claimed that when the distance from the observer to the visual stimuli was varied, corresponding variations in the visual angle subtended by the lights was irrelevant; the effect was always maximal when the lights were 7 cm apart.

In addition to these extraordinary findings, Scholz reported that analogous experiments on the tactile and auditory modalities yielded analogous results; apparent distance between stimulus sources was said to vary regularly with the temporal separation between the pulses, and the variation coincided with the temporal separations needed for illusory motion.

Neuhaus (1930) investigated the same question in the visual modality, but with somewhat altered conditions. In particular, Neuhaus doubted that motion could be seen well across 17° of visual angle, and so used smaller distances; he also objected to the lengthy stimulus durations Scholz used and to the procedure of direct comparison. In the outcome, Neuhaus claimed that he found no evidence of contraction, but only expansion; however, he also found that the magnitude of the expansion was correlated with the interstimulus interval, being maximal at the intervals that yielded optimal motion. In one experiment he even used unpracticed subjects who saw no apparent motion at the interstimulus intervals found favorable for practiced subjects, and measured a variation in perceived distance nevertheless; it was somewhat smaller than that found with the practiced subjects. Between them, Neuhaus and Scholz report variations in apparent separation of the flashed lines of as much as 60 per cent of their physical separation. This is an extraordinary range of variation.

Kolers and Touchstone (1965) therefore tried to measure this same effect, but with no success whatever. In their experiment a pair of lights was moved until it seemed to match the distance travelled by a light in illusory motion. Of three subjects tested, one showed a consistent overestimation of the distance that increased with an increase in interstimulus interval, a second showed a consistent underestimation that also increased with an increase in interstimulus interval, and the third showed no trend whatever. In other tests the marker lights were eliminated and subjects estimated numerically the distance travelled by the apparently moving object (magnitude estimation). There again one observer overestimated consistently, while three others underestimated the distance at short interstimulus intervals and overestimated it at longer ones. In neither experiment did the observers yield data that could be represented as a matching of apparent and physical distances at short and long interstimulus intervals (simultaneity and succession) and overestimation or underestimation at temporal separations between those extremes associated with optimal motion.

Puzzled by the discrepancy, Kolers and Touchstone conjectured that the results they obtained and, perhaps, the earlier results of Scholz and of Neuhaus, were accidental in a certain sense. They reasoned that in a totally dark room the eye, no matter how well fixated voluntarily, makes small perceptually uncompensated motions. In addition, a flash of light presented eccentrically induces the eye to make a reflexive motion that would normally bring the location of the flash onto the fovea, although the reflex is somewhat inhibited by the well-trained observer. These two motions of the eye could displace it sufficiently as to affect the retinal separation of the two flashes. The variations in judgment of distance then were associated with variations in position of the eye when it was stimulated, and not with the temporal or spatial characteristics of the flashes or with their special interaction in the nervous system. This conjecture was not proven (else it would not be a conjecture) and in any case cannot account for the results Scholz reported for tactile and auditory stimulation.

Related experiments on touch were carried out by two groups of workers. One effect was described by Helson and King (1931) for cutaneous perception and is called the tau effect. It is also obtained with visual stimulation (Geldreich, 1934). The other, a complement to it, is called the kappa effect (Cohen, Hansel, and Sylvester, 1953, 1955).

The tau effect consists in the report that the apparent distance between a pair of points varies with the temporal interval between them, as described earlier. For example if three points, *ABC*, are arranged so that the distance *AB* equals the distance *BC*, varying the temporal separation between *AB* and *BC* yields differences in their apparent spatial separation (tau effect). Turning the matter around to make the spatial separations physically unequal but the temporal separations equal, larger distances yield a perception of longer duration (kappa effect). Both of these phenomena, along with the results usually ascribed to the constant velocity hypothesis in apparent motion (Chapter 3) are probably based on the same temporal governors. No clear understanding is yet available of these many sorts of spatial and temporal interaction, however, and many variables affect the results (Marks, 1933; Takala, 1951; Fraisse, 1963, pp. 128–139; Cohen, 1964; Bill and Teft, 1969; Yoblick and Salvendy, 1970). It is not clear even how these many results should be interpreted in respect to the way the brain represents space, and the way the representations of space, time, and figure interact in the nervous system. What is clear is that a simple analogy to computer operation is false. Such an analogy assumes that there is a visual "program" in the nervous system that acts in an invariable way on the visual "data", much as a computer program acts on the data presented to it. The reason the analogy is false is that in visual perception, indeed, even in some cases of neural development, as Gaze, Keating, Székeley and Beazley (1970) have shown, the "data" presented to the nervous system affect the operation of the "program".

Figural Distortions

The interaction, whatever its form, is not simple. Brown and Voth (1937) predicted from their vector model of apparent motion (Chapter 4), that four lights arranged as the corners of a diamond would actually yield a perception of circular motion when they are alternated at the proper rate. Their theory predicted this, and they report observing it. Moreover, when only a single light was moved in true rotary motion, the diameter of its path seemed to be much less than the physical diameter, and to vary somewhat with rotation speed. The Brown–Voth contraction effect is much smaller than the effects reported by Scholz (whom they cite as providing substance for their prediction), averaging about 15 per cent at the maximum for illusory motion and somewhat less for real motion, but the magnitude is large still, and the phenomenon difficult to interpret easily.

The simplest response to make to difficult data is to deny their reliability and their authors' competence. The Brown and Voth experiment has often been commented upon along these lines, but despite the sometimes justified contumely, one or two facts resist dismissal. Sylvester (1960), for example, quite reasonably points out that four lights arranged as the corners of a diamond can as well yield parallel motion of the sides as circular motion, and reports that most of his observers did see the partial motions rather than the circular motion. These experiments were undertaken with free gaze; when a fixation point was added, some observers saw a single light tracing out a square, but five out of ten saw motion in a circle. (The perceptions are necessarily correlated with interstimulus interval: at very short interstimulus intervals the four corner lights appear on simultaneously, and at longer intervals they appear in clear succession.) Thus, under some conditions the Brown–Voth effect is realized.

About the same time that Brown and Voth published their results, Ansbacher (1938) gave the first report of an equally strange phenomenon in which motion affected figural size. A few years later, Ansbacher (1944) described his experiment in detail. In a partially lighted room observers fixated the center of a straight line while a 36° arc rotated around it in a circle whose radius was 20.7 cm. Both the straight line and the arc were lit from behind, and the speed of rotation of the arc was varied. The observer's task was to match the length of the straight line to the length of the arc, which was 13 cm long. (The chord of the 36° arc of this circle is only trivially shorter than the arc itself.) The judgments were made with the method of limits, the arc moving from faster to slower speeds in one set of trials and from slower to faster speeds in the other. Several results were obtained.

When the arc was stationary, the tendency was to underestimate its length, more if the arc was first seen stationary than if it was first seen moving. The underestimation varied considerably among individuals, but averaged to only a few per cent in the group. When the arc moved, however, extraordinary amounts of contraction were reported. At 30 revolutions per minute, the straight line matching the arc was set at

10 cm on the average, at 60 rpm it was set at 5.2 cm, and at 78 rpm it was set at 2.7 cm on the average. To put it another way, when the 13 cm long arc rotated at the subfusional rate of 78 rpm, its apparent length was matched by a line 2.7 cm long. In physical terms this is a contraction of about 80 per cent.

In control tests Ansbacher inquired whether the result was due to a dropping out of parts of the array or to a shrinkage of the whole, and concluded that it was shrinkage (or "telescoping", as he put it) rather than loss. When he varied the size of the arc, he found that the amount of shrinkage was not proportional to arc length but was a fixed quantity for given speeds (resembling Scholz's finding concerning visual angle).

In accounting for these impressive results, Ansbacher revived what has come to be known as the theory of the perceptual quantum or moment. In its recent forms (Stroud, 1956; White, 1963; Harter, 1967) the hypothesis has been put forth that the visual system accepts information not continuously but in packets or quanta between 50 and 100 msec long. Ansbacher's version of this old conjecture is based on the response of a hypothetical single retinal point that is stimulated "on" and "off" by a rapidly rotating arc. Having shown experimentally that a figure that passes its whole extent over a single retinal point undergoes more shrinkage than a line rotating through the same circle but drawn tangent to its perimeter, Ansbacher concluded that amount of overlap of contour at a retinal point is directly related to the amount of shrinkage. By means of some ingenious calculations he arrived at the value of 55 msec as the time-constant of the hypothetical retinal point, and thus of the pulsations characterizing visual perception. Moreover, on the basis of the theory of pulsations (its metaphorical analogue is that perception is based on the integration of a series of "snapshots") he concluded that illusory "stroboscopic motion is physiologically more elementary" than veridical motion (pp. 13–14).

More recently Marshall and Stanley (1964) and Stanley (1964, 1966, 1970) have examined various parameters affecting the shrinkage phenomenon. They confirm its existence and report that intensity and contrast affect its magnitude; also, that a dark arc on a lighted ground (opposite to Ansbacher's light arc on a dark ground) actually undergoes not contraction but expansion. They believe that both the shrinkage and expansion effects can be accommodated by the same mechanism, only slightly modified from the way Ansbacher proposed.

In arriving at his calculated value of 55 msec, however, Ansbacher took the apparent size of the stimulus as the measure of the true stimulus for a retinal point. For example, an arc 13 cm long that is 36° of a circle and that is rotating at 78 rpm, requires 77 msec to pass a retinal point. He argued that if the arc is perceived to be only 2.7 cm long, its passage over the retinal point lasts functionally for only 16 msec. Subtracting the 16 msec of apparent duration from the 77 msec of physical duration yields a "blind interval" of 61 msec at that retinal point, which is taken as a fair approximation to 55 msec. In this calculation, Ansbacher is treating the obtained distortion (the visual system's construction or "output") as a true measure of the applied stimulation (the input); this is an unrealistic use of the numbers. Hence his calculation of a time-constant is spurious, but his experimental results remain a challenge.

An interesting note on some of these effects, germane to the hypothesis of the temporal quantum, is found in an experiment by Lichtenstein, White, Siegfried, and Harter (1963). They presented a small number of pulses at a rate of 25 flashes per second to various regions of the eye and found that the subject's ability to estimate the number correctly remained fairly constant at different regions of the retina. Apparent duration of the pulse train, however, varied with the region of the eye stimulated. The eye of course is not homogeneous in its speed of report. A flash to the periphery takes longer to reach consciousness than a flash to the center (Sweet, 1953); hence a flash to a peripheral region of the retina must occur several milliseconds earlier than a flash to the center if the two are to appear simultaneous (Arden and Weale, 1954). The concept of simultaneity and the interpretation of such results are still not clear (Efron, 1967; Allport, 1968; Corwin and Boynton, 1968; Uttal, 1970); what Lichtenstein *et al.* have shown is that a single temporal quantum or psychological moment such as Ansbacher proposed cannot govern the eye's function, for at the least the "unit" of psychological time varies from place to place on the eye.

Clocks and Quanta

The concept of psychological moments is a perpetually interesting one to many psychologists nevertheless (see reviews by White, 1963;

Harter, 1967; Allport, 1968; Ansbacher, 1944, for some of the earlier references; and Moray, 1969, for its expression in the study of attention). Indeed, Stroud (1956) assumed that Wertheimer's interval of 60 msec for apparent motion was itself an index of the central periodicity of perception. Harter has performed the service of emphasizing the functional differences between the various models, and shows clearly how the recurrent interest is for many investigators related to the puzzle of cyclic brain waves. Despite its many forms the main proposal is that the visual system does not read its stimulation continuously but in time-based packets. In some models it is assumed that the peripheral receptor is blind intermittently (Ansbacher, 1944; Stroud, 1956), and in others that the peripheral receptor reports continuously but that its message is read or checked only intermittently (Harter, 1967).

The various aspects of the problem of accounting for intermittency have still to be sorted out. In some cases investigators seem to have confused the minimum time the visual system needs to operate on a particular input with an index of a driven rhythm, but that is faulty reasoning: the maximum speed at which a man can raise and lower his arm does not need to reveal the existence of a clock driving his arm; it may only reveal the amount of time it takes his nervous system to get the work done. Stimuli are not instantaneous in their effects nor is their encoding and operation instantaneous. The nervous system takes time to get the work done. Not clocks but encoding routines of finite and measurable duration may explain many of the perceptual processes now thought to reveal an intrinsic periodicity in the visual system. We shall see below that the concept of a clock is not needed to account for apparent motion although the opposite has often been thought. (Muijen, 1969, has recently reproposed rhythmical sampling as the mechanism.) The concept of a clock may not be needed either to account for the many inductions that are reported. (The medium of many of these interactions could be the path established by the assessor in Fig. 7.6, connecting the motion system and the figure system.) The phenomena, however, are still there to be studied, and important among those needing study are the effects of motion upon the perception of visual contours.

Other Modalities

These spatial and temporal interactions are obviously difficult to accommodate theoretically, but even they are not the limit of phenomena that may be called inductions. Apparent motion is reported not only visually, but in auditory and tactual space as well (Boring, 1942; Sherrick and Rogers, 1966; Sherrick, 1968). Sherrick has shown that a sense of movement can even be obtained between pulses presented separately to the two hands, despite the fact that the hands, unlike the eyes and ears, project each only to a single cerebral hemisphere, thereby demonstrating in another way that apparent motion does not depend upon a cortical short-circuit. Moreover, apparent motion is subject to intermodal influences. Werner and Zeitz (1928) found that two lights flashed a little too slowly to be seen as moving yielded an impression of movement nevertheless when their flashing was accompanied by rhythmic but slightly faster auditory or kinesthetic impulses. More dramatic still is the experiment of Galli (1932) recently repeated by Zapparoli and Reatto (1969). The latter authors first allowed observers to see, in one trial, visual apparent motion between a pair of lights and, in another trial, to hear auditory apparent motion between a pair of buzzers. Then in a third trial they simultaneously presented lights and buzzers and obtained what they describe as a fused motion of a luminous sound or audible light. Finally, in a fourth trial they presented a single buzzer and a single light with temporal and spatial separations between them. Their observers resolved the sequence as a motion between the two stimuli: "something that moves between the sound and the light or between the light and the sound, a light and sound tunnel which grows longer and shorter, or a light tunnel which grows longer and shorter while a sound passes through it" (p. 262). It is not exactly clear from this account what the observers experienced visually or auditorally. What is clear is that, in respect to the model suggested in Chapter 4, H-signals must be thought of as occurring intermodally as well. (The results of Zapparoli and Reatto suggest that the observer's ability to resolve disparate V-signals is virtually limitless.) Perhaps because of the transmission times characterizing the response to stimulation, the nervous system constructs motion-like perceptions from stimuli either in the same modality or in different modalities. Each modality of course has its own special way of transducing stimuli; the occurrence of inter-modal apparent motion raises several problems of interpretation.

One interpretation suggested by the data is that the perceptual system performs similar operations in its different departments; a perceptual experience need depend not upon a difference in operation, but upon a difference in the information on which or the place in which a particular operation is carried out.[1] The nervous system seems to resolve temporal disparities in very similar ways, irrespective of where they occur. Of these ways, the most prevalent is a distortion or transformation of the perceived object. The perceptual representation is of plastic and modi-fiable objects, formed at the behest of suitably asynchronous signals, whose appearance depends partly upon the signals themselves and partly upon the contents of the dictionary of options (Fig. 7.6).

A question raised by the results asks whether individual modalities are special cases of a master modality, so that the disparate stimuli are processed directly, or whether intermodal phenomena require abstraction and comparison of information from the separate modalities. The answer remains to be seen (and heard). The variety of induced per-ceptions appears at the least to create substantial challenges for theory. Of all the inductions known, however, perhaps the most impressive is the occurrence of apparent motion itself. In the next chapter some of the subject-dependent aspects of the phenomenon, another form of induction, are discussed.

---

[1] Von Békésy (1967, 1969) has described a number of ways in which analogous results are obtained from different modalities when analogous stimulating con-ditions are used. The old phenomena of intersensory influences (London, 1954), moreover, find their modern counterpart in comparisons of speed of judgment of "sameness" and "difference" regarding stimuli presented intramodally and inter-modally (Posner, 1970; Sternberg, 1970).

CHAPTER 10

ATTITUDES, SKILLS, AND PRACTICE

ABSTRACT
The third kind of induction effect is due to the subject himself. The effect of practice in seeing the illusion is to enhance its clarity and distinctiveness, but to restrict the range of intervals over which it is reported. The perceptual experience is rather immune to the influence of volition, but it is sensitive to the influence of special skills and knowledge. The chief role of experience is to create perceptual capabilities (perceptual options) within the observer which increase with practice. Pictorial and linguistic experience have separate internal representations ("dictionaries"); these sometimes make contact and sometimes do not. Not all pictorial experiences are encoded in words, nor do all words have pictorial counterparts.

There are a number of aspects to the effects of practice on the perception of illusory motion. One is the role of repetition of the stimuli upon the ability to see apparent motion; others are the roles of set, attitude, and volition, and of special skills and knowledge. The role of repetition is two-edged, for in some conditions it lessens and in others it enhances the perception of motion. Let us first consider the inhibitions.

Inhibitory Effect of Repetition

The major inhibitory effect of repetition is to induce a breakdown of the perception. DeSilva (1928) obtained his subjects' reports after each of a very large number of pairs of flashes within a single testing session; he found that the likelihood of seeing motion and the quality of the motion seen declined with successive trials. To put it another way, the "threshold" for seeing motion changed during the course of the testing session, a well-known effect in psychophysical measurements. In a related experiment Kolers (1964) also found that the perception of motion will break down when two lights are alternated for a lengthy period at equal interstimulus and intercycle intervals. The perception of motion is replaced by two lights flickering in place; but then, as with other ambiguous figures, the first perception returns, only to lapse again. Flicker and motion thereafter alternate in perception.

Both of these observations show that some neural conditions essential to the perception of motion change with time, but whether as a matter of "fatigue" or some other process is not known. One way to study the possibilities is to try to track the motion perception over the period of observation. In my experiment the interstimulus and intercycle intervals were kept constant while the variations in perception were reported. One could do the experiment in the opposite way: varying the interstimulus and intercycle intervals in an effort to maintain the perception. It would be interesting to find what one has to do to the conditions of stimulation—duration of the flashes or of the intervals between them—in order to maintain the perception for prolonged periods, and even whether one can maintain the perception for prolonged periods at all. Alternation of the perception between two states bears so striking a resemblance to the data obtained with other ambiguous stimuli whose perceptions alternate, such as the Necker Cube and other geometric illusions (Cohen, 1959; Axelrod and Thompson, 1962; Kolers, 1964) as to make most worthwhile a careful analysis of the conditions affecting these many ambiguous figures. It is a reasonable conjecture that some of the alternations are induced by a common mechanism. On the other hand the perception that a figure is continuously present while its motion comes and goes suggests in another way that the figural and motion components of the illusion are relatively independent.

Facilitative Effect of Practice

I have described two ways in which repetition inhibits the perception of motion; there are many ways in which it facilitates it. Two such effects of repetition have already been mentioned: Wertheimer (1912) found that after repeated pairings of the flashes, the observer continued to see motion for a few additional trials despite sudden elimination of one of the stimuli. A second effect is that the spatial separation over which motion can be seen is increased by practice (Fig. 2.2). Indeed although it is still not known what the spatial limits on the perception of good motion are, the tests made with a cathode-ray tube, illustrated by lines *a* and *b* in Fig. 3.8, found good apparent motion reported over a distance of about seven degrees of visual angle. Those tests were made with highly practiced subjects, and that range is larger than the analogous effects reported by Neuhaus for two flashes of light (Fig. 3.1), where a limit seemed to be reached at about four degrees of visual angle, and by the movement of an arrow enclosed within a rectangle (Fig. 4.7), which also seemed limited to about four degrees of angle. In sum then, one effect of repetition as expressed by extended practice may be on the spatial range over which apparent motion can be seen.

A more obvious positive effect of repetition is on the clarity of the percept. Although vision, like language, is creative in allowing for new events to be perceived, the clarity and interpretability of the new perception is poor at first, and only increases with familiarity. Both linguistic clichés and familiar shapes are perceived far more readily than novelties. DeSilva (1929, p. 284) remarked that "First reports on unexpected apparent movement situations indicate that the visual data were unclear and inadequate. After repetition, the visual data became more adequate. The observers experienced greater confidence and were able to report definitely and with a fair degree of accuracy." We have confirmed the truth of this description in our own tests. In one case, temporal conditions for optimal motion were found when the intercycle interval was 3.2 sec. One of the two cards used for the display was then changed during one intercycle interval. On the first two or three trials thereafter, the quality of the motion and the clarity of definition of the moving object were seriously impaired, but then recovered with repeated trials. A distinct sense of motion was always present to the subject after the change of cards, but identification of the moving object and the clarity of its definition were poor. Recent work by Haber and Hershenson (1965) has confirmed these well-known effects of repetition, although with another kind of display.

In observations of this kind it is the clarity of the moving object that is affected by repetition. In other tests, we find that the precision of the observer's response, as expressed by the temporal range of interstimulus intervals yielding positive reports, is also affected. Practice at the task

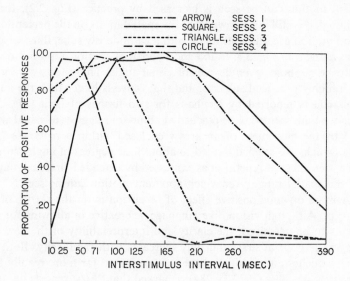

Fɪɢ. 10.1. The likelihood of seeing motion between four different pairs of identical shapes in four successive testing sessions. Data from a single subject.

created by repetition induces the subject to refine the categories of his response.

Pomerantz and I studied some aspects of the subjects' acquisition of this skill. Experimentally naive subjects were shown alternating flashes at various rates and durations until they spontaneously reported a perception of motion. They were then aided in refining their criteria for the perception, and subsequently were tested with different identical pairs of shapes. On each of four test days the subjects made 600 observations with a single pair of shapes: a pair of squares, a pair of triangles, a pair of circles, or a pair of arrows. Days of practice and sequence of test shapes were arranged in a 4 × 4 latin square; thus, each subject saw a unique sequence of pairs across the four days, but the sequences were counterbalanced across the four subjects.

The shapes were always presented for 80 msec each, the interstimulus interval was varied, and after each trial the subject reported during a 3.2 second intercycle interval whether smooth, continuous motion had

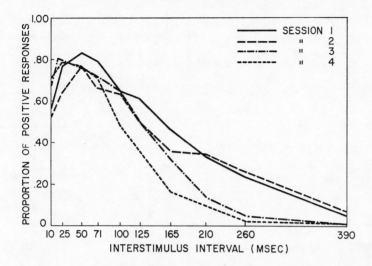

FIG. 10.2. When the sequence of pairs is counterbalanced across subjects, the effects of practice (sessions) still remains strong.

been seen. Figure 10.1 shows the results for a subject who made the observations with arrows on the first day, squares on the second, triangles on the third, and circles on the fourth. Notice the marked shift in the curves along the abscissa. The area under the curve for arrows is several times greater than for circles. To what is this due?

The main question is whether the shift is due to the difference between the pairs of shapes presented, or to increasing practice at the task. The answer is shown in Fig. 10.2, which is based on the data for all four subjects, collected with a counterbalanced order of sequence of shapes. The figure shows that the difference is due not to the shapes themselves, but to practice at the task, for despite the counterbalanced order, a strong effect of practice remains (here shown as testing sessions). A consequence of practice is that the range of interstimulus intervals at which motion is reported becomes smaller. Positive reports at longer interstimulus intervals especially are fewer, for the subjects come more reliably to distinguish smooth continuous motion from a discontinuous but alternating sequence.

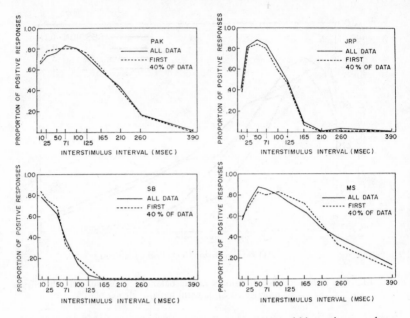

FIG. 10.3. The changes in criteria apparently do not occur within sessions, as shown by the high correlation between the first 40% of each subject's data and all of his data.

We may note, parenthetically, that other experimenters have also found greater variance in the data for the distinction between motion and succession than between simultaneity and motion (Brenner, 1953). A satisfactory sense of motion can be attained despite the absence of figural continuity. H-signals seem to dissipate slowly and the judgment of succession is correspondingly more variable than the judgment of simultaneity. This slow dissipation, and the distinction between motion signals and figure signals, may explain the conditions sometimes called "inference of motion" (Chapter 3).

Having found these changes of criteria, we ask when they occur. On each test day the subjects made 600 scorable observations, in 10 blocks of 60 each. Did their criteria shift during the course of the testing, because of variations in attention, set, or alertness? Two analyses of the data reveal negative answers to this question.

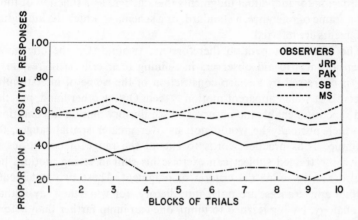

Fig. 10.4. A further demonstration that criteria remain stable within sessions.

In one analysis, the first 40 per cent of the subjects' responses during a test block were compared with their responses for the whole block; that is, the results after 24 observations at each of 10 interstimulus intervals were compared with the results after 60 observations. Figure 10.3 shows those comparisons. The subjects differ individually in their thresholds and range for seeing apparent motion, but the curves for 40 per cent and 100 per cent of the data are quite close to each other in every case.

In the second analysis we calculated the percentage of positive responses the subjects made in each of the 10 test blocks on each of the four days. Figure 10.4 plots the percentage of positive responses in each of the blocks, summed across days. (The percentage within each block therefore is based on 60 observations × 4 days = 240 observations by each subject.) The curves show that the subjects differed in their criteria for reporting apparent motion, but that the criteria remained stable within a test day. The conclusion seems to be clear that the observed variations in criteria are due to events that occur between testing sessions and not within them. Repetition alone does not necessarily change the criteria; a lengthy pause between testing sessions seems to be required for those changes to occur. Presumably it is not variation in criterion for individual judgments that changes as a function of time, but a frame of reference, a standard, or a schema to which the individual judgments are referred.

The effects of repetition therefore are twofold. On the one hand, repetition acts to aid observers in refining their criteria of response; this is expressed by a sharp constriction of the range of interstimulus intervals at which apparent motion is seen. On the other hand, practice and repetition enable the observer to see illusory motion in conditions in which normally he would not, as over greater spatial ranges, for example. Thus practice works to extend the range of conditions in which the trained subject can exercise his skill in seeing motion, but it also acts to refine the criteria of judgment. These two statements, despite appearances, are not contradictory, for what they say is that, by analogy, having learned to jump, one can jump farther than before, but having learned to jump, one can distinguish more precisely between jumping and walking. Having acquired a visual skill, one can extend the range of its exercise, but having acquired the skill, one can better distinguish it from other skills. Perceptual learning is not only a matter of refining discriminative capabilities (E. J. Gibson, 1969); it is also a matter of applying the skills in novel situations. Both of these phenomena seem to be matters better described by perceptual schemata than by sensory analysis.

Attitudes and Volition

The notion that schemata of judgment are affected in the perception of motion might imply to some that suggestion, prestige, or volition are the basis of the perception. Illusory motion, like most psychophysical phenomena, is quite sensitive to practice and experience. Indeed sometimes it is asserted that this illusion especially is only a matter of suggestion. Kelly (1935) exposed a pair of flashing lights to a classroom of students and found that only about half of them reported the appearance of motion spontaneously. He then described what could be seen, whereupon 90 per cent reported seeing motion on a second trial. Was it "suggestion" or "prestige" or "influence" on the experimenter's part that enabled the students to see the phenomenon, or did the description merely direct the students' attention to some perceptual possibilities? In answer, one can say that if the conditions for illusory motion are not favorable, suggestion and prestige notwithstanding, the subject will not report seeing it (as long as he is faithful to his perceptual experience and is not merely lying to please the interrogator). Suggestion and description can facilitate having the perception under certain conditions, as can expectation and attitude as well, but none of these is its cause or explanation; indeed, seeing the illusion cannot be brought entirely under voluntary control. Three observations bear on this point.

In one test (Kolers, 1963) two small disks were first alternated at a rate that yielded the appearance of smooth, oscillating motion. A few vertical lines were then placed in the empty field during the interstimulus and intercycle intervals, their horizontal separations slightly larger than the diameter of the disks. The subjects were instructed to will seeing sinuous motion of a disk between the upright lines, but were unable to do so. Hence mere intention or willingness to see a particular effect did not enable its occurrence.[1]

In a second experiment Kolers and Pomerantz (1971) alternated two trapezoids, one of which was a 180° planar rotation of the other (R of Fig. 6.1). Subjects could sometimes see one trapezoid rotate in the plane to become the other, but the prevailing perception was a combined inversion and mirror reflection: the first trapezoid seemed to come out of the plane of the display and simultaneously rotate on its horizontal (inversion) and vertical (mirror reflection) axes. This complex motion was neither expected nor could it be completely inhibited by the subjects.

The third observation was described earlier and has a similar implication. Two lines of light oscillated at a rate that yields good apparent motion will after a while be seen to be flickering. The motion experience is temporarily lost, but soon returns, only to disappear again. The two percepts, flicker and motion, alternate with each other, much as the two phases of a Necker Cube or other ambiguous figure alternate. When subjects are instructed to will seeing one or the other orientation of a cube, they can affect the reversal rate of the ambiguous figure somewhat

[1] This result is in marked disagreement with one reported by Koffka (1931). Taking issue with an earlier demonstration of Linke's, Koffka flashed a pair of disks and an arc and reported that a disk could, by exercise of volition, be seen to roll within the arc. I have not been able to repeat Koffka's or Linke's observations.

(Bruner, Postman, and Mosteller, 1950). In like manner subjects instructed to will seeing motion when the lights appear to be flickering, or flicker when they appear to be in motion, can also affect the durations of the cycles somewhat; but neither perception can be maintained exclusively. No matter how hard the trained subject wills seeing motion, he loses that perception after some moments of observing. The phenomenon cannot be brought entirely under voluntary control; indeed, the effect of volition is small at best. The phenomenon of illusory motion is as natural a consequence of the phased presentation of stimuli as is any other psychophysical effect the consequence of its stimulation, and may be susceptible to extraneous influences in a way that is only quantitatively different from that found with other psychophysical judgments.

Schemata or frames of reference may be constructed by people to cope with individual situations; but they also exist and act as general cognitive attitudes that are taken toward phenomena. Investigators discovered very early that some attitudes facilitated seeing apparent motion whereas others inhibited the perception (DeSilva, 1926). In general it was found that a quiet, passive, object-oriented attitude helped, whereas an active, analytical, stimulus-oriented attitude interfered with the phenomenon. I have often found, for example, that engineers especially find it difficult at first to see the illusion, particularly when they know that it is produced by the flashing of only two lights. They sometimes think that seeing motion from flashing lights represents a "distortion of reality", and they resent the implication that their senses are "playing tricks" on them. (Engineers and physicians may form the largest membership in the contemporary group of naive realists.) With encouragement to relax and be permissive toward their experiences, they usually can manage to see it, however.

The emphasis on a quiet, passive attitude suggests immediately that certain cerebral states may underlie the distinction between analytic and passive attitudes. The analytic attitude is usually associated with a high degree of alertness, and variations in states of alertness are now known to be associated with levels of activation of some of the lower regions of the brain. If variations in level of activity of the limbic system are indeed associated with these variations of attitude, an interesting puzzle is raised as to the means by which signals from the brain stem and limbic system modulate signals in the visual network; that is, how does "attention" affect perception functionally? The effect is quite general. For example, as mentioned earlier, an image that is made immobile on the retina by means of suitable optical devices will soon have its perceptual representation fade out and finally disappear (Riggs, Ratliff, Cornsweet, and Cornsweet, 1953). Using a telescope arrangement (Pritchard, 1961) I have had subjects report on the disappearance. After the disappearance, a sharp flick to the subject's cheek, neck, thigh, or calf would result in the immediate reappearance of the shape, which then promptly disappeared again. One might think that the flick to the cheek induced a momentary displacement of the telescope relative to the eye, especially if the subject moved his head sharply in response; but even if this were the case, a flick to the thigh or calf would be unlikely to elicit the same reflexive movement. One might conjecture however that the flick induced a short-lived change in the state of the subject's "level of arousal", which in turn affected the level of his visual functioning (Worden, 1966). Variations in level of arousal may therefore be the basis of the difference induced by "analytic" and "phenomenological" attitudes. This need not be the only basis of the difference, of course, for with level of arousal more or less steady, the nature of the attitude itself may be important to the perception. Judgmental attitudes ("I like it", or "It is good") may themselves structure perceptions differently from the way analytic attitudes ("It is composed of such and such") structure them.

Interpretability of the flashes (DeSilva, 1926; Toch and Ittelson, 1956), and expectations about them, sometimes are reported to affect the measured temporal characteristics of apparent motion. The main effect is on the range of interstimulus intervals at which naïve subjects report that motion can be seen. Sometimes that range is a little larger when the flashes are semantically rich than when they are simpler geometric shapes, but the method of measurement may even affect the influence of expectations and interpretations. Jeeves and Bruner (1956) found that expectations about direction of motion increased the interstimulus interval at which subjects would report motion in an ascending method of limits, but had no effect on the values measured using the descending method. As in Brenner's (1953) report, here too we find that subjects are apparently willing to continue seeing alternation as motion once they have seen motion; but initiating the response sequence with clear instances of succession, expectation seems to play little role in the conditions first called motion. The difference between motion and succession is difficult to measure accurately, and is the region where the sharpest tightening of criteria occurs, as shown in Fig. 10.2. The role of practice on such effects has not however been evaluated.

That expectation or interpretability affect the perception at all raises some issues for understanding the mechanism of the phenomenon. One question concerns the time at which the perceptual representation of motion occurs. Wertheimer's theory implicitly, and van der Waals and Roelofs' (1930) theory explicitly suggest that the motion percept is delayed in its construction until the second of the two flashes is registered in the nervous system. Van der Waals and Roelofs presented two flashes of different color or of different shape and found that the motion between them was nevertheless smooth and continuous. They argued that the visual system would not "know" how to transform the stimuli until the second one had appeared; hence the entire percept, in their view, must be delayed in its construction until after the second flash has been presented and registered in consciousness.

This apparently plausible hypothesis can be rejected for several reasons. One is that in most experiments on apparent motion the subject is shown the same pair of stimuli on each of a number of trials. [Indeed, Neuhaus (1930) even reports that he did not begin his measurements until after showing the subject the array about 20 times.] Good information about the nature of the flashes is conveyed by repeating them. Hence the subject's knowledge and expectations about the content of the flashes can provide him with sufficient information at the beginning of a trial to initiate the construction of the motion perception before the second flash is presented. To put it another way, the motion percept can be thought of as generated in real time, not in delayed time. Another reason to reject the hypothesis is Wertheimer's observation that motion can be seen even in the absence of a second flash; but further discussion of this subject will be delayed until Chaper 11.

Here again we may distinguish the mechanisms active in the construction of the motion aspects of the perception from the mechanisms operating on the figural interpretation. Repetition and expectation affect clarity and convincingness of the perception; but a global or diffuse sense of motion is experienced even when the flashes are as radically disparate as a photograph of a dining room and the face of an Indian. Using the latter pair with experienced subjects, we found that initial flashes are reported as diffusely or globally moving; with repetition, a few features of one display grow into features of the other and the motion becomes more smooth and compelling, much as DeSilva (1926) described it. The conditions underlying the illusion are not affected to any notable degree by volition, attitude, or expectation; the observer's interpretation of "the visual data" can be affected, however. Hence we may expect that certain skills and forms of knowledge may affect some aspects of illusory motions.

Knowledge and Special Skills

The assumption of a pure visual state and of the intrinsic virtue of working with naive subjects are both false beliefs. Describable visual experiences inevitably incorporate skills and information in naming, identifying, categorizing, and discriminating, whether the stimuli are real scenes, pictures, or psychophysical stimuli (Gombrich, 1961; Goodman, 1968; E. J. Gibson, 1969). A subject unpracticed in a particular perceptual task is not a visually naive subject; his lack of skill in the task at hand yields data having greater variance and less reliability than those obtained from a person skilled at the task. His data do not reveal the operation of a "pure" visual system.

Miles (1931) reported many years ago that observers tend to interpret ambiguous or illusory motions by direct analogy to familiar mechanical movements. These "naive" subjects behaved thus despite the fact that in so doing they were not being faithful to all the signals from the retina. They responded in terms of familiar categories of experience, and not with fidelity to the stimulus situation. Greater experience and more skill would have enabled them to be more particular, giving more detailed and richer reports.

In some tests we used markedly disparate shapes, such as an arrow and a small circle, or a trapezoid and an arrow, and our subjects were topologists and geometers or clerks and typists. The mathematicians would occasionally describe elaborate or esoteric transformations of the shapes that sometimes could and sometimes could not be seen by non-mathematicians, especially not by clerks and typists (nor even by the experimenters). For example, most people see an arrow and a circle undergoing plastic deformation in planar motion. However, one geometer spontaneously reported the array to be in depth, the arrow swinging through the third dimension to appear front-on (the circle). Apprised of this possibility, other subjects could sometimes then see that transformation. Analogous results are obtained with other ambiguous transformations such as that of Fig. 5.3, Array 21. An observer does not always spontaneously report seeing all of its possibilities. Encouraged to look for one in particular, he will sometimes report seeing it after engaging in some sort of subjective willing, expecting, or intending. We are ignorant of the mechanisms of this interaction between linguistic instruction and pictorial representation, but the phenomena are readily demonstrated.

One cannot always see what other people see, however. One topologist we tested reported seeing transformations of the trapezoids of Fig. 6.1 that even after his elaborate verbal description Pomerantz and I could not persuade ourselves that we could see. Hence it is not the verbal information or suggestion by themselves that induce the interactions mentioned. The visual apparatus contains its own dictionary of pictorial possibilities which the dictionary of linguistic descriptions can sometimes contact. When this occurs verbal description can facilitate pictorial representation. But the two dictionaries are not identical. One may contain linguistic categories with no pictorial equivalent; and the other may contain pictorial representations having limited correspondences in language. The topologist whose experience and, perhaps, special skills induce him to see certain transformations can sometimes describe them to other people who, themselves, can then sometimes see them. The gifted art critic, historian, philosopher, and scientist do the same, conveying by linguistic means the structure and content of their own perceptions of pattern. In a sense, we use language to teach each other how to see what our visual apparatus has all along been looking at and registering. It seems implausible to me that linguistically induced variations in perception operate directly on the processing of stimuli at the graphemic level; but it seems equally implausible that the influence is exerted only at the level of conscious interpretation. An extremely difficult problem is the means by which these two systems we use for representing the world interact so that our pictorial experiences can drive us to invent new linguistic descriptions, and our linguistic experiences induce us to perceive new pictures. It may be a formidable problem to explain how words can induce a person to see pictures in a particular way, but it may not be any more complicated than explaining how people see pictures at all.

Two assertions are obvious regarding these interactions between linguistic and pictorial representations. One is that the visual system operates according to its own rules of transformation. No one we tested expected to see a planar rotation of trapezoids transform in the third dimension (R of Fig. 6.1), or see depth from neighboring flashes (Figs. 5.5 and 5.6) but these were the spontaneous responses from our observers. The second assertion is that what the visual system makes of stimulation depends somewhat upon what the viewer knows. If the first statement means that certain conditions of stimulation are necessarily resolved as particular pictorial representations, the second statement means that most of our visual stimulation is actually equivocal. There is no escaping the perceptual consequences of a flash of light presented to the intact open eye. Every eye responds much the same way, and irrespective of what name is given to the flash, it is a flash that is seen. Remarkably few stimuli of the natural world are so unequivocal. By far the greater number have physical characteristics that fall within a common range of intensities, sizes, wavelengths, and the like. What we make of those depends as much upon what we are as it does upon what they are. Attitudes, experience, and skills mediate the interpretation of what is seen. Enriching the viewer's vocabulary of possible pictorial representations by inculcating skill in their creation, makes such items available to his visual constructions. Moreover, as with words, presumably also with pictures: the more familiar and well-used a particular item, the more readily it can be generated internally. One need not think, as some do, that familiar visual items are recognized by means of a template or by holistic pattern-matching; speed of retrieval may be increased, and the clues needed decreased, with practice.

Practice is important to seeing apparent motion (Kelly, 1935; Raskin, 1969) but, as with any psychophysical procedure, practice alone is not its explanation. The unpracticed subject may be unwilling to describe as smooth motion the meager one-third of a second sequence of two disparate shapes; but if after much practice he can so describe it, we cannot say that this is due to suggestion, compliance, or delusion. We therefore are faced with a classical problem in perception research. Does one take as a mark of the visual system's characteristics what the unpracticed system or the practiced system reports? If the unskilled observer cannot perform, are we to say that the visual system does not usually or typically perform in this way? Clearly not. Moreover, if the practiced subject can perform a task, can we say that only one method is used by the nervous system to achieve this end? Also clearly not. Hence our research must study what the unpracticed subject *can* learn to do and what the skilled subject *has* learned to do. The performance of neither subject provides unqualified insights into the function of the perceptual apparatus.

These remarks have a bearing on the description of apparent motion. The classical division of the illusion is into five stages: simultaneity, phi motion, partial motion, optimal motion, and succession. These five can easily be doubled or trebled, however, when the skilled observer pays attention to small details. He may see "jitter", a slight oscillation between simultaneity and partial motion; he can distinguish continuous movement from a brief dwell in place prior to motion when flash duration is varied; he can distinguish global motion from clear figural change when the shapes are disparate; and sometimes he can distinguish smooth continuous change from replacement.

The list can be extended, but to no point. As Neff (1936) suggested, what the Gestaltists described as stages are only parts of the whole. Partial motion is not a stage toward which the visual system works; it is an incomplete visual construction. A fuller description of the illusory construction would pay attention to more of its attributes as part of a continuity and not only to those the Gestaltists selected out and treated as distinctive categories of experience. Neff (p. 22) cites a work by Wittman in which not three or five, but eight stages are described; to which it is easy to reply, why eight and not eighteen? The possible number is actually boundless. We must therefore always try to distinguish the experimenter's convenience in describing from the perceiver's actual operations. Both reflect the influence of practice, skill, and knowledge, for the experimenter no more has a pure mind than his subject has a pure eye. The skilled observer establishes categories that name increasingly finer distinctions between perceptual experiences, as the scientist himself does in the interplay between his theory and his observations. When in our daily affairs the distinctions do not make a difference, we say that things are similar or look alike. With practice we learn to distinguish differences, thereby lessening the variance of our data. An uncertainty in the study of perception, however, is whether these skills come into play only at the level of interpretation, or whether the visual system's operations are affected from the earliest stages on by existential characteristics of the observer, not only in what he chooses to attend to, but in the way his nervous system represents that information pictorially. In any case, when the need arises we can always tell a difference out to the limits of our acuities and our cognitions. Insufficient appreciation of the facts cited has led to much misunderstanding of the role of practice and of naive subjects; consequently, of the perceptual equivalence of real and illusory motion.

The Equivocal Equivalence of Motion Perceptions

A very great number of investigators have maintained that real and illusory motion are perceptually equivalent. For Wertheimer the equivalence had enormous significance because of its implications for his theory; indeed, the quotation from Wertheimer in Chapter 2 states explicitly that, as perceptual experiences of motion, there is no difference whatever between real and illusory motion. Even a non-Gestaltist such as Gibson (1954, p. 310) has said that it is "unfortunate" that a difference was ever drawn between the two, a sentiment shared by many others.

The notion of equivalence has at least two important aspects. One is the experiential equivalence or perceptual similarity of real and illusory motion; the other is the mechanism proposed for the two perceptions. As Hartmann (1935, p. 7) remarked: "Since Wertheimer claims that as far as perceptual experience goes there is no difference whatsoever between the perception of real and illusory movements, he is implicitly proposing that wherever two identical phenomena are found, it is necessary to assume that the corresponding brain-processes are identical." Obviously if one can demonstrate important perceptual differences between real and illusory motion it follows that the corresponding brain processes cannot be identical.

Most judgments of equivalence were based on memory. Indeed it is the case that two lights flashed at the appropriate rate remind one of or look like an object in real motion so far as the appearance of motion goes. DeSilva (1929) however made direct comparisons of real and illusory motions, relying on observation rather than memory, and thereby began to place important restrictions on the conditions of equivalence. Anything other than optimal motion, he found, such as partial movements or succession, was always readily distinguished from real motion; and distinctions could also be made if the movements were slow (about 3° per second), or if they extended through the longer distances he employed (about 3.2°). Greatest confusability of real and apparent motion occurred in the range of speeds from 10° to 21° per second at the smaller spatial separations of the flashes (less than 3°). In other words, some conditions make it more difficult and some make it less difficult for the observer to distinguish between veridical and illusory motion, but distinctions can be made. (Parenthetically, DeSilva disproved "suggestion" as an explanation of the effect. If suggestion were the cause or principal basis of the phenomenon, its influence should not have been tied so tightly to stimulus conditions.) But what does "equivalence", even in restricted conditions, mean? All it means is that optimal motion "looks like" real motion. However, if the two events did not "look like" each other in some respect, there would be no basis for calling one of them an illusion of motion.

Similarity of appearance is a trivial basis on which to build strong arguments. Speaking strictly, nothing looks like anything else, for no matter how carefully crafted or cunningly contrived a second object or event is, it can always be told from another by some means; and that difference pointed out, the "similarity" can disappear. Contrariwise, anything can be found to have some similarity to anything else.

Museum curators are frequently faced with the problem that forgery creates: to tell apart two objects that look alike in content, style, or method of fabrication. Goodman (1968) has illuminated some of the epistemological aspects of this problem. In brief, judgments of similarity are always made in respect to some set of criteria that are taken as standards. Hence, "like" and "unlike" are as much aspects of judgment as aspects of stimuli. Every object or event has an infinite number of characteristics; the skilled looker learns to seek a few distinguishing features that enable him to tell apart objects or events that to a novice (i.e., "naïve subject") appear the same. Such distinguishing marks can always be found if one has the wit, the means, and the patience. The point that two things look alike begs the question because it uses the sloppiness and imprecision of normal experience (Bartlett, 1932) as an index of the capabilities of the visual apparatus. Although it is useful to have information about how the visual system can be misled because of its biases and proclivities, and evidence that some features are more likely to be responded to than others, this information only describes but does not explain the system's functioning. The resemblance of real and apparent motion is a problem to be explained, not an answer to a question.

The distinction between the two perceptions does not necessarily require judgment and reflection; the visual system itself can make it. On the basis of studies of visual masking with stationary stimuli, Kolers (1963) reasoned that a line of light in real motion should inhibit the perceptibility of a dim target light in its path, a conjecture subsequently found to be true (Luria and Kolers, 1962). I wondered whether, in light of the similarity of appearance of real and illusory motion, a similar inhibition of appearance of a target would occur with a line in illusory motion. The results were quite straightforward: despite the similarity of appearance of real and illusory motion, a line of light seen in illusory motion did not affect the perceptibility of objects in its path. (The variance in the data was sometimes affected, however.) The perceptual equivalence notwithstanding, the visual system constructed the object that was seen to be moving in ways sufficiently different as to achieve two different experimental outcomes. Thus the perceptual equivalence does not support an argument for functional or operational equivalence.

Experiments on animals have been used to bolster the argument for equivalence. Several authors have reported that rates of intermittency that yield good apparent motion in man also induce responses similar to those found with real motions in insects, frogs, guinea pigs, and cats. The responses measured were behavioral (Smith, 1941; Rock, Tauber, and Heller, 1965) and electrophysiological (Grüsser-Cornehls, Grüsser, and Bullock, 1963). The findings have been used to argue for the "innateness" of the apparent motion response and for the functional equivalence of veridical and illusory motion. The data cannot be used to support either argument, however. What they may support is the idea that H-signals play an important role in mediating a motion response, and may have similarities across species. The mechanism cannot be exactly the same in all species, of course, for the anatomy of the eye and details of its operation differ among them. The insect has a compound eye, which neither cat nor man has; and movement detectors are found in the retina of the rabbit (Barlow and Hill, 1963), but only in the cortex of the cat (Hubel and Wiesel, 1962). The fact that intermittent flashes of light stimulate responses normally associated with moving objects tells us something about a basic or necessary condition that characterizes the construction of apparent motion. But they tell us little more than that, analogously, flour and water are involved in the construction of items as disparate as bread, pastries, and wedding cakes. Neither the operations nor their outcome are the same in all cases. Analytical experiments can reveal some of the constituent operations.[2]

Another assumption built into the assertion of equivalence concerns the argument from appearance to function. Because two things look alike in some respects, does it follow that the nervous system constructs their representations in identical ways? Consider an analogy from mathematics: Four sticks may individually be counted and the quantity "four" be held in mind. The numbers 3 and 1 can be added to yield 4, or one may take the square root of 16, or the logarithm of 10,000. In each case the quantity held in mind is the same, but the particular operations generating it are quite different, for they are as varied as enumeration, summation, root extraction, and exponentiation. The identity of results can never be used as an argument for the identity of processes. The occurrence of visual illusions suggests in turn that the limitations on visual encoding are categorial not functional: The number of possible perceptual experiences is infinite, but the number of defined experiences is small. Hence, many different routes through the nervous system yield outcomes that are treated as "same". Saying that illusory motion looks like real motion means that we have not found it necessary to establish

[2] The epistemological banality of difference in appearance does not help to illuminate this issue. However, we might be able to find that a cat or a man can learn to make different responses to optimal illusory motion and veridical motion, thereby finessing the question whether optimal motion looks like real motion to a cat. One might conjecture that an alert and well-trained man would reveal a narrower range of equivalence conditions than frog or cat would.

categories of perceptual experience that distinguish the two. We may assert on theoretical grounds, therefore, that it should always be possible for a skilled observer to tell apart illusory and veridical motion on the basis of some perceptual criterion. Practice and skill are important components in these judgments, and their manipulation as empirical variables in research on perception could illuminate many issues that today are dark and confused.

CHAPTER 11

THEORIES OF APPARENT MOTION

Finally, we come to examine in detail some of the theories proposed to account for apparent motion. Three kinds of theories have been put forward, but none of them is found adequate. Some problems for theory created by the empirical observations are discussed, such as modality stimulated and the role of consciousness. Contrary to most other assertions, here it is argued that the illusion is probably created in "real time".

In the preceding chapters I have discussed many of the experimentally derived facts of apparent motion. Let us turn now to some of the theories associated with them. Although there are many differences among the theories, they can be grouped into three comfortable classes, Figural theories, Excitation theories, and Epiphenomenal theories. The first two kinds attempt to explain the phenomenon, and the last attempts to explain it away. I begin with the last therefore in order to show that the phenomenon, subtle, fragile, and curious as it is, cannot easily be explained away.

Epiphenomenal Theories

Many of the epiphenomenal theories take the form they do because their authors do not fully distinguish between an influence and a mechanism. A number of conditions influence perceived motion, such as eye movements, body tension, and attitude, as the preceding three chapters have shown, but the discoverers of these influences are sometimes insufficiently self-critical when they suggest that the influence is the mechanism. Perhaps chief among the epiphenomenal theories is the one that in various forms attributes the entire experience to an involuntary movement of the eye during the interstimulus interval.

Eye motions can indeed influence perceived motion, as Pomerantz

(1970) showed; it was even suggested in Chapter 8 that involuntary motions of the eye may account for some of the variations in perceived distance reported between the flashing lights. Higginson (1926) identified some other characteristics of the phenomenon that he thought could be attributed to eye motions; the most important is the frequent perception that the illusory object appears from its onset to be in movement. Pomerantz and I found, however, that with short durations of the first flash, less say than about 150 msec, the illusory object seems to appear from beyond the screen and to move across it to its terminus; but with longer durations of the first flash the object appears in its first position, seems to dwell there for a moment, and then to move to its terminus. Higginson's argument presumably would be that the first perception is due to a brief, involuntary, eye movement, which is compensated for when the stimulus has a longer duration; but that does not explain the illusion itself.

Higginson began his note in a rather diffident tone, but by its end had claimed that eye movements were a complete and parsimonious explanation of all illusory motions. The claims stimulated Guilford and Helson (1929) into undertaking their elaborate disproof of Higginson's thesis, photographing their eyes during the observers' perception of motion. They concluded that not only is there no positive correlation between the two, but that eye movements sometimes seemed to interfere with the attainment of optimal motion.

In light of modern knowledge of the variety and subtlety of eye movements (Cornsweet, 1956; Westheimer and Mitchell, 1969), Guilford and Helson's work probably does not prove their case, if only because their apparatus could not resolve small motions of the eye. Their case was actually proven even before they undertook their investigation, however, as they well knew. The disproof of eye motions as a condition necessary for illusory motion has two parts: one is that analogous motions do not occur with auditory or tactile stimulation, yet illusory motion is also perceived in those modalities. The second, as they also point out, is that split motion and motion in opposite directions were well-known effects—they had been reported even by Wertheimer (1912)—and a single eye obviously cannot move in opposite directions simultaneously. (No one seems to have suggested that the retina or eyeball stretches to accommodate motion in opposite directions, for which we may be thankful.) As a third disproof, if it were needed, one might add that plastic deformation, apparent change of orientation, and a perception of depth from unequal interstimulus and intercycle intervals can hardly even be conceptualized as due to eye motions let alone be explained by them. Eye motions influence, but are not the explanation of illusory motions.

The method of their influence is, regrettably, not clear. One wonders whether eye motions affect only the interpretation of the moving object, such as its perceived direction of rotation, or whether they may induce excitations that themselves affect the pattern of signals obtained from the retina. The latter question has to do with the "cancellation" and "suppression" theories, or "inflow" and "outflow" theories of visual motion perception. These deal with the fact that there is an asymmetry of effect between optical image and retina. A stationary eye and moving object yields a perception of a moving object, but a moving eye and stationary object typically does not. The means by which the visual system learns or knows whether the relative displacement between image and retina is due to object motion or to eye motion is an old and difficult problem in perception, and is under intense investigation at the present time (MacKay, 1962; Stoper, 1967; Matin, Matin, and Pola, 1970). Indeed, it is just because motion signals dominate shape signals that the visual system must have some means of taking motions of the eye into account, for otherwise the visual world would be figurally unstable. A clear understanding of these processes is not yet available, however.

Several other epiphenomenal theories in the older literature emphasize subtle kinesthetic influences of various kinds. Kinesthetic strain was an important component in much of Wundt's and Titchener's theorizing and some investigators influenced by Wundt used these notions to try to account for apparent motion, but without much success. Neff (1936) has reviewed much of this earlier literature.

Another group of epiphenomenal theories that dismisses illusory motion might be called Equivalence Theories. The main thrust of these works, as I mentioned in the preceding chapter, is based on an observation that has been reported by almost all investigators of apparent motion, that optimal motion is perceptually indistinguishable from real motion. The conclusion drawn from this observation is, all other things aside, that the identity of appearance implies an identity of mechanism.

Wertheimer (1912), who had a theory of apparent motion, draws this inference, and Gibson (1968) and Gregory (1966), who do not have a theory of apparent motion, draw the same inference, as have many other students in between (for example, Aarons, 1964; Morinaga, Noguchi, and Yokoi, 1966). Gibson once remarked that is was "unfortunate" that a distinction had been drawn between veridical and illusory motions; Gregory offers the metaphor of a loosely fitting lock and key to account for the equivalence; and others take the equivalence of appearance as evidence of identity of mechanism. Of course, the argument from equivalence of appearance to identity of mechanism cannot be sustained logically or epistemologically, as was pointed out in Chapter 10; moreover little has been done that actually tries to define the conditions of equivalence. DeSilva (1928) began some research along these lines, manipulating only temporal intervals, and showed that the equivalence existed only for a limited range of temporal and spatial conditions. The work needs extension. How bare and stark must the stimulus array be before the subtle and experienced observer can no longer distinguish between real and apparent movement? What clues are useful in aiding the distinction? What spatial separations, if any, make a distinction impossible? Knowing that one is seeing an illusion does not affect the threshold for motion (Segal and Barr, 1969) and willing has little effect upon the perception. It would be worth learning what the limits of apparent equivalence are.

Beyond these distinctions lie some formidable facts that prove that the theory of identical mechanisms is invalid. The theory assumes that stimulating the retinal space between the flashes is irrelevant to the perception of motion, and that only sequence and timing of the flashes is important. Hence, the intervening stimulation is thought of as redundant. As I reported in Chapter 3, however, stimulating the intervening space does not give rise to a monotonically improving perception of motion. Two separated flashes yielded a satisfactory perception of motion, and 32 or more flashes within the same spatial extent also yielded good perceptions of motion, but four or eight flashes did not. The minimal conclusion must be, therefore, that the intervening flashes are not redundant, for at certain temporal and spatial separations they do something to the encoding of motion that alters the perception qualitatively.

A second fact invalidating the thesis of identity is our persistent inability to obtain the perception that two illusory objects cross paths (Chapter 5). This failure is not attributable to distance between the flashes nor to their timing, but what it is that inhibits the perception of crossed illusory motion is not yet known. Crossed paths of motion can be seen with physically moving stimuli, of course.

There are some other ways that veridical and illusory motion differ. Illusory motion obtained from two flashing lights does not yield a motion aftereffect (Humphrey and Springbett, 1946), whereas motion from a moving contour does. Moreover, the continued inspection of beta motion needed for an aftereffect to develop, induces beta motion to break down instead. The physical stimuli seem to flicker in place, but no motion is seen between them for some interval; suddenly the appearance of motion returns, only to disappear again. The alternation between motion and flicker represents, as it were, two states of an ambiguous perception. The perception of depth from flashes whose interstimulus and intercycle intervals are unequal (Chapter 7) further distinguishes the two perceptions of motion, for the analogous perception with real motion would be change of rate, not depth. And finally the figural distortions of plastic deformation and change of orientation are produced in ways that must necessarily be different from those that characterize the perception of contours in real motion.

We can account for some but not all of these differences. Suppose that, as was earlier suggested (Kolers, 1963), visual inputs are analyzed as a sequence of stages, so that at each stage something new is added to or removed from the product of the preceding stage. Suppose too that the stimuli for apparent motion are "artificial" in the sense that they share some but not all of the characteristics of the stimuli for real movement. The artificial stimuli would pass through several early and coarse stages of analysis, as so many forgeries pass careless scrutiny, but lacking the proper ingredients or features, would not have done to them and would not do at later stages what the veridical stimulus does. This is approximately what Gregory (1966) had in mind with his analogy to a lock and key. The trouble with this idea is that some perceptions of illusory motion are dependent upon the objects flashed, but others are not, so that in general the analogy is much too coarse to be of any real use. The theory of identity tends to assume that there is only one way to achieve a motion perception, which I have shown is not the case, and that illusory and veridical motion achieve that outcome in similar ways, which is also wrong.

Although the argument has long been made, the basis for asserting the identity of mechanism of real and apparent motion is weak. In the long run we shall probably find it conceptually simpler to note some similarities of appearance between the phenomena and some substantial differences in mechanism. We know little enough about how the visual system encodes real motion; we certainly know far too little for it to be useful to ourselves to "explain" illusory motion, about which we know even less, as an instance of real motion.

Figural Theories

One set of theories that tries to explain the illusion of motion is quite cognitive in its orientation. Their main thrust is that the visual system resolves a disparity in perceived locations by creating a sense of motion. In some early forms of the theory, figural identity was assumed to be important. It was thought that the observer first saw a figure in one location, then saw the same figure in another location, and by noting the identity of the figures and the disparity of the locations, was induced to create the perception that a single figure had moved. Strongly emphasized by Linke and by Hillebrand (Neff, 1936), this emphasis upon the cognitive component has persisted in various forms to the present time (DeSilva, 1929; Rock and Ebenholtz, 1962; Stoper, 1963; Muijen, 1969).

The early theories that emphasized figural identity were refuted by the many experiments carried out with disparate figures. If any two shapes can deform plastically into each other, identification of the same object in two locations is obviously not requisite to the motion perception. Hence the strong form of the figural identity theory is not correct. But even a weaker form is not correct. Suppose that a notion of figural similarity is substituted for figural identity so that, say, a triangle is said to be more similar to a square than to an arrow (or vice-versa). The difficulty here is that the notion of "similarity" itself is quite difficult to interpret (Goodman, 1970). Efforts to metricize visual shapes along dimensions of similarity have not been rewarded with much success (Kolers, 1970). Indeed, the perceptions of motion obtained with disparate figures even suggest that all plane figures are functionally equivalent in the visual system at a certain level of operation.

Any two shapes that change into each other as easily as each moves with its likeness must be objects that the visual system treats as members of the same class. The phenomenon of "replacement" reveals that not all displays will change into each other; but many simple plane figures such as circles, triangles, trapezoids, and irregular polygons will. The shapes that do transform are different at the level of "surface" or graphemic features; but these features are not fundamental to figure construction, or plastic deformation would not occur between them. The sensitivity to contour revealed by the occurrence of replacement must be defined at a more abstract level than can be accommodated by a shape's perimeter, angularity, area, or similar features. Moreover, as Kolers and Pomerantz (1971) pointed out, "strengthening" a contour by increasing the duration for which it is exposed actually increases the likelihood that the "strengthened" shape will undergo transformation. Contour, in the narrow, specific sense of that term is not important to the analysis preceding apparent motion; the level at which it becomes important remains unknown. Contours, therefore, cannot be a fundamental aspect of visual processing but a relatively late one, else strengthening a contour would retard its perceptual alteration. Theories of pattern recognition that regard contours at the graphemic level as the fundamental unit of perception, for example, Sutherland (1968) and Dodwell (1970), are insufficiently abstract in this regard. The aspects of figures that are important to the visual system's processing are probably not best identified with the measurable aspects of their geometry. The lack of success the metricizers of shape have encountered may be due to their trying to distinguish between shapes all of which, so far as the visual system's operations are concerned, are members of the same class of object. Hence, even the weak form of the figural identity theory is of severely limited utility in accounting for apparent motion.

There can be no question, however, that some observer-dependent characteristics of figure, such as skill at manipulating the stimulus and interpretability of the disparity, affect the clarity and convincingness of the perception of an object in illusory motion. Even with displays as rudimentary as those Pomerantz and I used, repetition of the array led to a clearer and more sharply defined sense of motion. What is at issue with these theories, therefore, is not whether familiarity or meaningfulness influences the clarity and convincingness of the perception, for they unquestionably do; rather at issue is whether a well-formed visual perception of a figure in disparate locations is the principal requirement for the illusory motion. Here the answer is negative.

A small effect can almost always be found, so that some pairs of flashes will not have exactly the same smoothness or clarity of motion as other pairs. What this comes down to is that figure exerts an influence on the interpretation of the perception of motion, but is not a cause of the perception. Frames of reference and spatial characteristics of the display may influence thresholds and interpretation of the motion experience (Cohen, 1964); hence, the perception resolves the disparity between the stimuli, but a perception of their disparity is not the basis of the phenomenon.

The proponents of a cognitive theory, moreover, have the difficult task of explaining why the perception has its unique temporal characteristics. Why can an object and its copy be seen in two locations at very brief separations (simultaneity) and at long ones (succession), but not at the intervening intervals giving rise to motion? This question has often been answered by resorting to another cognitive component, the clearing-up time between successive figures (for example, DeSilva, 1929). It is argued that the visual system does not have enough time to clear up consciousness between the flashes, hence moves one figure into the other instead. This however is merely a restatement of the main thesis that the conscious perception of figures is essential to the illusion. If this were the case, figureless phi-motion would not be seen, but of course it is. Contrary to these theories, we may assert that the visual system does not create illusory motion by first identifying a shape in different locations and then adding motion to resolve the disparity. Rather, the system responds to the H-signals and moves the shapes generating them, whatever they may be. A consequence of this assertion is the idea that the construction of figure in apparent motion is substantially different from the construction of figure in veridical contour perception. This topic will be pursued in the next chapter.

Excitation Theories

Both the figural theories and the excitation theories have characteristically distinguished between the perception of an object and the perception of its location; indeed, the study of motion perception has historically been concerned with the interrelation between these two aspects of perception. How difficult that interrelation is to conceptualize is freshly illustrated by recent research on fish and rodents that seems to suggest that different parts of the brain may do the processing for the two components (Ingle, 1968; Schneider, 1969). But where the figural theories emphasize the perception of figure, the excitation theories emphasize the disparity of locations and the consequences of stimulation at disparate sites.

Probably the best known of the excitation theories is that of Werthemer (1912) and Köhler (1923). As I have said several times now, they proposed that the separately stimulated retinal loci generate regions of excitation in corresponding areas of cortex which, when properly timed, interact with each other electrically. The main advantage of the classic Gestalt theory is that it proposes a particular rather than a general metaphor to account for the U-shaped function of apparent motion and clearing up time: simultaneity is the analog of stimulation whose two parts occur too close in time for the excited regions to interact, succession is the analog of stimulation whose parts are too separated in time, and optimal motion is the analog of proper timing. The evidence against this form of the theory is overwhelming: stimulating heteronymous retinas yields good motion (Smith, 1948); the cortex itself is not needed for some kinds of motion perception (Smith, 1941); the theory does not account for the perception of depth from disparate figures (Neuhaus, 1930); it cannot explain why depth is perceived when other objects are flashed in the path of illusory motion (Chapter 7); and it certainly cannot accommodate the difference in perception between plastic deformation and depth (Kolers and Pomerantz, 1971). The short-circuit theory has been refuted perhaps more times than any other in perceptual psychology, yet it must capture a quality that many investigators find attractive, for it has lingered on.

The main variants on the short-circuit theory are the vector model of Brown and Voth, and the magnetic model that was tested in Chapter 4.

Although the vector model may have predicted the results Brown and Voth reported, those results are not always found (Sylvester, 1960). Moreover, the model predicts 45° vector resolution for two displays placed at 90° angles, but this is not observed (Fig. 5.3, Array 17). We may conclude that the vector model, and its underlying idea of "attraction", does little more than apply some concepts of mechanics uncritically to visual perception.

The mechanism mediating apparent motion could be wholly independent of figure—a compulsory response to flashes of proper distance, energy, and timing. If this were the case and the motion experience represented the action of such an innate factor, then motion could readily be seen between any pair of stimuli. Alternatively, the mechanism could be wholly dependent upon figure. If this were the case, motion could be seen only between identical figures. The facts support neither view.

What we find, rather, is a system in which appropriate flashes induce a compulsory tendency to a perception of motion, but one whose figural properties are monitored from the outset. The adequacy of the figural monitoring contributes to the clarity or distinctiveness of the motion perception, to the occurrence of depth rather than plastic deformation, to Orlansky's fusion, and to the phenomenon of replacement. Figure has been shown in our experiments to be subservient to the compulsion to motion, but not altogether without influence. Hence, any system proposed for apparent motion must take account of the interaction of what for brevity I have called H-signals and V-signals. Purely excitational theories, although they are closer to the facts than figural theories, fail in this important respect.

Other Problems

No theory of illusory motion has yet been proposed that accommodates more than a few observations. Two reasons may lie at the heart of this failure. One is that apparent motion is not an anatomically localized phenomenon. Wertheimer (1912) and the Gestaltists, finding that the phenomenon could be obtained when the flashes were presented one to each eye (dichoptic stimulation), argued that the phenomenon was "central" rather than "peripheral", and not only central but cortical.

No such conclusion is warranted. Dichoptic stimulation can be used only to make a negative case: failure to obtain a percept dichoptically proves only that most of the processing occurs early in the visual channel; but the occurrence of a percept dichoptically never proves the reverse, that all of its processing is normally cortical. To be sure, both visual masking of contours and the perception of illusory motion reveal notable differences in their frequency and amplitude of occurrence as a function of mode of observation; characteristically, the effects are stronger when both flashes are presented to the same eye than to different eyes. Moreover, Gengerelli (1948) and Shipley, Kinney, and King (1945) have shown that faced with the choice of seeing motion from signals within one cerebral hemisphere or between two, the visual system creates motion from signals in the same hemisphere somewhat more readily. However, the fact that illusory motion can be obtained when stimuli are presented separately to the two hands (Sherrick, 1968), each of which is represented in only one-half of the brain, and between different modalities (Zapparoli and Reato, 1969), resists explanation altogether in terms of local cortical interactions.

Rather than exclusively here or exclusively there, what seems to be the case is that the motion experience can be generated in a number of places, and from a variety of stimuli. Motion can be perceived when the two flashes are presented to the same eye, either on the same or on opposite sides of the fovea; when the two flashes are presented dichoptically to homonymous or heteronymous retinas; and when the stimuli are presented to the ears, the hands, or even intermodally. The perception of motion seems to be a primitive way in which spatially disparate excitations are resolved by the nervous system. Moreoever, it seems to be an effect in which the "same" perceptual result is produced by operations carried out in different places in the nervous system; hence its occurrence cannot be identified with "motion detectors" of the visual system. The strength of the perceptual result may vary with the place at which the interaction first occurs, but the place only rarely determines the occurence of the phenomenon itself. The richness of perceptual experience is due not to a richness of operations that the nervous system performs but to the variety of signals on which it exercises its limited repertoire.

A second reason no fully adequate theory has been proposed may be that the role of time in the phenomenon has not been clarified. The question to be answered is why the phenomenon has the time course it does. Many writers who reject the Gestalt theory of a build-up of cortical excitations have turned to notions of cortical scanning rhythms, neural quanta, temporal gates, central clocks, and the like. In earlier days, when timing apparatus was less flexible than now, investigators were sometimes restricted to one or another temporal value in their investigations. The tradition grew up, not yet entirely ended (Sgro, 1963), of finding *the* interstimulus interval at which apparent motion occurs. In point of fact, there is no single interval, but timing is important nevertheless. As I showed above, motion can occur between signals associated with onsets and between signals associated with offsets. The important point is that there be transients in the visual system that may interact.

As either the leading or trailing edge of a flash may provide the needed H-signal, the sum of intervals from the onset of the first flash to the onset of the second does not fully accommodate the facts of timing. In any case, this sum, as Neuhaus (1930) pointed out, is not itself readily interpretable because a 10 msec flash plus a 90 msec interstimulus interval gives very different results from the reverse. The interstimulus interval can even be reduced to zero (as some early investigators showed) and motion still be seen, but obviously will not if the flash is reduced to zero. Hence, onset to onset time is asymmetric, and is at best only a rule of thumb guide to the results.

Another issue concerned with timing asks when the visual processing occurs. Because the perceptual experience rationalizes the disparity between two presentations, several writers have implied that the visual system must wait until the second stimulus has appeared before it undertakes the construction of the motion perception.

Implicit here is the idea that the conscious representation of the first and second events is prerequisite to the motion relation between them. This view however confuses perceptual representation with perceptual construction. The greatest part of our processing of stimuli does not require their conscious representation, for much of the work goes on as the silent operation of the machinery. The perceptual construction represents the results of stimulus processing, not the means by which stimuli are processed. (Neisser, 1967, has reviewed some of the evidence supporting these contentions.) There is little reason to believe that conscious perception or registration of the second flash must precede construction of the motion experience in a single trial.

To the contrary, the model suggested in Chapter 4 implies that the timing of signals accounts for the construction. The model explicitly denies a delay in the perceptual construction, and affirms that it occurs in the "real time" that characterizes all visual processing. We know that pictorial representations follow the application of stimulation by about a quarter to a third of a second. The figural resolution is a later stage in processing than the integration of motion signals, but the construction is carried out in as "real" a time as is, say, backward masking.

Curiously enough, although many workers have asserted that the construction is delayed, none has undertaken to measure the putative delay itself. That should prove easy to do. Let four different reaction times be measured, the speed with which a subject can press a key on the appearance of an object. The four measurements are of: (A) the speed of reaction to the first flash when it is presented alone; (B) the speed of reaction to the first flash when it is followed at various intervals by the second; (C) the speed of reaction to the second flash when it is presented alone; and (D) the speed of reaction to the second flash when it is preceded at various intervals by the first. These may be symbolized as (A) $RT\ 1|1$; (B) $RT\ 1|(1+2)$; (C) $RT\ 2|2$; and (D) $RT\ 2|(1+2)$. Let all measurements of RT be made from the onset of the first flash, a blank card substituted for one stimulus figure in (A) and (C) above. The hypothesis of a delay in perceptual construction must assert that at interstimulus intervals that yield apparent motion, $RT\ 2|(1+2)$ is significantly longer than $RT\ 2|2$. RT to the second flash alone should be a simple linear function of the interstimulus interval whereas, if the assertion is correct, $RT\ 2|(1+2)$ must be linear for simultaneity and succession but be concave downward at interstimulus intervals between. The hypothesis of real time processing denies the latter assertion. (Measurements of reaction time to the first stimulus are control conditions for reliability.)

In doing the experiment one would want to guard against the subject's anticipating the occurrence of a stimulus; rhythmicity of occurrence has a powerful effect on speed of reaction (Klemmer, 1957,

1967). Varying the interstimulus intervals in a random manner from trial to trial should be adequate as a control. In assessing the results, moveover, one would want to take into account the potentiating and inhibiting effects on reaction time of the pairing of flashes (Smith, 1967). With these considerations in mind, the experiment could be a direct test of the question of timing, and be carried out for phi motion, plastic deformation, and rotation in depth of rigid shapes. The hypothesis advanced here is that the reaction times will reveal no exceptional delays in visual processing of apparent motion, although the subtlety of the discrimination may affect the variances of the data.

For reasons that are still not fully clear, the visual apparatus does not characteristically analyze in detail stimuli that arrive within a brief interval of each other. A full understanding of the limitations on visual resolution is itself difficult to come by. The difficulty lies with apparently contradictory facts: A flash presented 250 msec after another can interfere with the perception of the first (backward masking), two flashes separated by 100 msec appear as a single object in motion, and two flashes separated by 30 msec appear simultaneous. These results reveal poor temporal analysis. Nevertheless, as von Békésy (1969) has shown, the visual system under certain circumstances is sensitive to differences in time as short as 0.1 msec. In addition, Mayzner and his colleagues (Mayzner and Tresselt, 1970) have found that temporal intervals of as little as 40 microseconds can make a difference to some visual functions. Thus on the one hand we find much evidence that the visual system characteristically does not function as a temporal analyzer but rather as a temporal integrator; and on the other hand, we find evidence that the visual system is sensitive to remarkably small variations in timing, which speaks for temporal analysis (compare Chapter 7, footnote 2). There is a slight suggestion in this evidence that the difference lies with the fact that the system integrates stimuli that are redundant for energy, location, or content, but distinguishes or differentiates stimuli of certain disparities or phase differences; but this suggestion will need many qualifications even if it is correct in principle.

Although the phenomena are not well understood, their range actually argues strongly against clocks, quanta, gates, and the like; against discontinuities in visual processing and for the idea of continuous sampling. Otherwise, variations in duration of the flashes would not have the effects they do. Timing cannot be avoided conceptually in dealing with apparent motion. In the model sketched in Chapter 4, however, timing has to do, not with clearing up in consciousness or with temporal quanta or gates, but only with the intervals required for signals to be transmitted, and the duration for which stimulated locations remain active. Timing is important in this model only as a consequence of short-lived excitations; the excitations are not timed but they last a length of time. A clock does not control the visual system, but only measures the duration of its activity.

CHAPTER 12

RECAPITULATION

ABSTRACT
The two themes running through this work are the notion of "a language of vision" and the nature of perceptual experience. The argument now made is that visual patterns as such are not amenable to classification into the types required for treatment in a language; all shapes seem to be members of a single class. These results are pessimistic for theories of pattern recognition that are based on graphemic differences between shapes. As to perceptual experience, the argument is made that the nervous system represents two different kinds of information when we have a pictorial experience, externally supplied and self-supplied, and does so according to different rules for the two.

Two themes run through the work I have reported. One is the possibility of establishing a notational system for pictures, a device that would rid the psychophysicist of his embarrassing difficulty in specifying his stimuli (other than by name or measurement) and that would enable people to "write" pictures to each other much as engineers can write circuits, musicians can write sounds, and poets can write language.

The problem has not always been well-stated nor stated thus, but it comes down to establishing a "language" for visual perception. An older enthusiasm produced such titles as "The language of vision"; more recently we hear talk of the "grammar of vision" and "the psycholinguistics of cinema" (Pryluck and Snow, 1967). The idea behind this is the implicit belief that visual perceptions are constructed by a serial ordering mechanism in such a way that a picture is put together in the mind as, so to speak, a visual sentence. The goal therefore is to specify the visual analog of syntax and of parts of speech, the rules by which the bits and pieces of visual information gleaned from the retinal image are put together into a pictorial whole, and the units upon which the construction proceeds.

Some familiar experiences of equivalence of shapes (Dodwell, 1970) support this enterprise at one level. We classify many different shapes together, such as all kinds of circle-like objects. The visual system treats many objects as categorial equivalents, but we are ignorant of how such categories are formed. Are the category memberships based on names? If that were the case, speechless people and language-less animals would have narrower bounds on their equivalence relations than people with language have. The evidence from ethology and from developmental psychology suggests the opposite, however: language itself acts to narrow the bounds on equivalence categories. Perhaps then the equivalence relations are based directly on visual aspects of objects? If this were the case, we should be able to establish perhaps the alphabet, perhaps even the words of the language of vision, by obtaining judgments of similarity between shapes and measuring the shapes to find the basis of those judgments.

Establishing the units of such a system is the implicit aim of the shape metricizers, and the explicit aim of the work reported above on disparate shapes. I had hoped that if the visual system were confronted with large numbers of markedly disparate shapes, it would, by the ease or difficulty with which it created a perception of motion, tell us in an empirical way which shapes belonged together and which were different, thereby aiding us in establishing the "units" upon which our perceptions are built. The results indicate that shapes indeed are not all the same, but differences among them are revealed more by their detail than by overall aspects of contour. Replacement and figural fusion occur when the fine structure of two shapes is dense and disparate, but not as a function of their contours. In studies of visual masking, it was once pointed out (Kolers, 1962), the visual system seems to operate separately upon the contour and upon the interior of shapes, revealing a U-shaped function for the masking of high-contrast contour and a monotonic function for the masking of interiors. A somewhat similar distinction seems to be operative in the figural aspect of apparent motion: any two contours can be seen in motion but the occurrence of figural fusion and replacement seems to depend more upon the detail enclosed by them than upon the contours themselves. In any case, the hope of establishing a notational system for shapes on the basis of the ease with which the visual system can transform them into each other was not realized in the data obtained. The visual system seems to be able to transform any two contours into each other.

The implication of this finding is a pessimistic one for psychophysicists; it is that shapes, and pictures generally, are not capable of being further notationalized than by name and measurement. It may be an inherent characteristic of the visual system that it can regard any shape as representing any object, so that the very plasticity of visual interpretive mechanisms (the plasticity that resolves V-signals, for example) precludes the development of an efficient notational system for shapes.

Some years ago Lewin (1931) distinguished between phenotypic and genotypic descriptions of objects, between their measurable surface features and the rule to which they all responded. All objects that fall, for example, can be separately measured and described, but their falling is itself captured by a "higher-order" principle of gravitational attraction. A similar difference between levels of description is sometimes expressed in linguistics as (the difference) between the surface structure and deep structure of a sentence, that is, between the many ways a relation can be expressed and a minimal symbolic representation of the relation. The analogous situation in the present work is that many shapes that differ in their measurable surface features all readily transform into each other when the shapes are alternated. The distinctions of geometry that separate them at the level of their surface features are not important to their mutual transformation: in Lewin's sense, the phenotypic differences are submerged in their genotypic similarity. For the visual system responding to their alternation, all of the shapes are members of a single class. Hence, establishing figural units of perception at the level of graphemic analysis of shapes seems to be proved impossible by these results.

It is possible, of course, that the critical differences that distinguish shapes are important only at later stages of visual processing than those tapped by the phenomena studied here. It is possible that impletions occur at a stage where the visual processing is concerned only with gross features of the stimuli. Hence, one might find that the system paid more attention to differences among shapes if one could make the measurements farther along in the visual construction. Plasticity of shape might be a characteristic only of a stage of global or schematic representation and not therefore the whole story. At later stages, shapes might be found to be more resistive to distortion and transformation.

The problem here is to specify a level of description that is narrow enough to make it meaningful to talk about visual operations upon classes of shapes, and yet broad enough for more than one shape or family of shapes to be included in a class. If all two-dimensional shapes are members of the same class, as they seem almost to be shown to be in the present results; or if distinctions are drawn only at the limits of detectable differences between shapes, as the uniqueness of pictures implies they may be (Ivins, 1968; Goodman, 1968), then the idea of establishing classes of shapes according to the visual operations performed upon them is hopeless. Shapes may be categorized in many ways, but the categories are conceptual, not "visual". To put it another way, shapes and their features that we see and describe are the results of visual processing, the output of visual productive mechanisms, but not the units or elements of processing.

The second theme running through this research concerns the nature of perceptual experience. Perceptual experience has two sources, "information" supplied by the environment and "information" supplied by the perceiver. No perceptual experience is entirely free of transformational operations performed by the perceiver, even if the operations are only those of photochemical response. Many perceptual experiences are free of external sources of stimulation, however. In the standard psychophysical experiment the perceiver is used as a detection device, and the experimenter does all he can to reduce the "noise" of the measurements due to fluctuations of threshold, interpretations, change of attitude, and the like. The ratio of environmentally-supplied to self-supplied information in the resulting pictorial experience comes as close to $1:0$ as is ever possible with a conscious subject. In dreams, voluntary images, and related conditions, the pictorial experience is created by the perceiver in the absence of the physical referents of his perception. Here, the ratio of environmentally-supplied to self-supplied information comes close to $0:1$. Dreams and voluntary images can be as compelling pictorially as representations of stimuli that are physically present. At least since Perky's (1910) experiment we have known, furthermore, that under proper circumstances the perceiver cannot tell from the pictorial experience alone whether what he is seeing is based on environmentally-supplied or self-supplied information (Segal and Gordon, 1969). Even though the perceiver does not know what the source of his pictorial experience is, we may ask an empiricial question: does the visual system itself make a distinction in the operations it performs in constructing these pictorially equivalent experiences?

The illusion of motion stands somewhere near the middle of the hypothetical dimension upon which we can place sources of pictorial experience. The illusion is keenly responsive to the characteristics of the physically supplied stimuli at the same time that its occurrence requires active supplementations by the perceiver. Hence, the ratio of sources of information in this illusion stands as something like 1:1. On the basis of the experiments carried out on the illusion, it seems clear that the ways the visual system uses the two sources of information are markedly different in certain respects. To help make this point it is necessary to discuss the theory of feature analysis and to contrast it with pictorial impletions.

Two Modes of Representation

The main metaphor now used to describe figure perception is feature-analysis. This notion maintains, contrary to the assertion of the Gestalt-ists, that perceptual experience is built up over time out of bits and pieces of information extracted from the image on the retina. The extraction may be described in psychological terms as an analysis of the image into its major distinctive figural aspects that, once extracted, unmistakable identify the object (Werner, 1935; E. J. Gibson, 1969); or such analysis into distinctive features may be followed by algebraic recombination of the parts into a whole (Sutherland, 1968). The extraction may also be described in physiological terms as the activity of "tuned" analyzers or detectors that respond selectively to attributes of the stimulus such as its orientation, motion, density, and the like (Sekuler and Pantle, 1967; Gilinsky and Doherty, 1969; Pantle and Sekuler, 1969; Blakemore and Campbell, 1969; Campbell and Maffei, 1970). Much of this work finds its motivation in the discoveries made during the past forty years in electrophysiological investigations of single cells of the visual system, but especially in the more recent discoveries of motion and edge detectors, and of complex and hypercomplex figural analyzers

(Hartline, 1934; Kuffler, 1953; Lettvin, Maturana, McCullough, and Pitts, 1959; Ratliff, 1961; Hubel and Wiesel, 1965).

The physiological evidence shows that the visual system is capable of carrying out a great deal of analysis of features of stimuli. The method of operation is that the deeper into the visual system one goes, the more selective a cell's response to stimuli is. Retinal ganglion cells will respond to any excitation in a general region of the eye, cells of the lateral geniculate bodies respond more narrowly, and cortical cells in the striate and peristriate areas repond more narrowly still. This selectivity of response is the strongest evidence we have for feature-analytic operations in the encoding of visual stimuli. The notion proposes that the image on the retina is, as it were, "digested" selectively in the pathways of the visual system in a manner somewhat similar to the "analysis" of nutrients that occurs when food passes through the alimentary canal. The food is met in its passage by localized enzymes, acids, and other chemicals that respond selectively or interact with the mass. "Features" of the mass are "recognized" by the localized chemicals and various aspects of the mass are operated on in various regions of the canal, sugars here, proteins there, and fluids in another place. Digestion is a feature-analytic process with some of it, moreover, occurring serially (as when one kind of decomposition must occur if another is to occur) and some of it in parallel. The result of this decomposition is of course the transfer from the canal to the interior body of the elements or units required for its many functions, the substances and energy that enable us to survive. The corresponding transfer from visual analysis is "information". What I have shown is that, as with a body that feeds on its stores and thereby supplies itself with its own energy, the visual system supplies itself with its own information.

A digestive model of analysis is entirely consistent with the idea that figure and motion are analyzed as separable components of stimulation. It makes it questionable, however, whether that distinction should be described as "two visual systems" (Schneider, 1969) rather than as two analytical operations.

The visual system reveals clearly marked selectivity in its response to veridical stimuli, as documented above. I shall show now that these biases do not affect the figural constructions of impletion. The evidence comes from three sources.

1. When a figure such as a Necker Cube or one of Gottschaldt's elaborations is alternated with a part of itself, the visual system does not perform the logically simple operation of analyzing the common part out of the complex and move just the part. Rather, it grows the lines or the square into a whole cube in one direction of motion, or shrinks the cube into the part in the other direction of motion. The physical presentation is represented at the termini of movement, for the observer sees the whole cube or its presented parts there, but the intervening stages are synthetic supplementations, not analytic decompositions.

2. Increasing the duration of a flash typically increases the likelihood of its correct report and description of its features. Hence, increasing flash duration should permit better analysis of the figure. Increasing the duration of the first of two disparate flashes, however, increases the likelihood that it will lose its features; the longer its duration, the more smoothly it transforms itself into another shape. Thus, increasing the opportunity for feature analysis actually increases the likelihood of shape change (loss of features).

3. In psychophysical and electrophysiological investigations, orientation of lines, grids, edges, and shapes is found to have an important effect on their perceptibility. The same characteristics are found to play almost no role in supplementations of apparent motion. Indeed, Frisby (1968) found that similarly oriented and differently oriented lines and contours go equally well into motion, revealing that the "detectors" implicated in studies of veridical contours and veridical motion play little if any role in the perceptual transformation and elaboration of figures in apparent motion.[1] The experiments of Kolers and Pomerantz (1971) similarly reveal little sign that feature analysis is important in the subjective supplementations of apparent motion. Although veridical presentations may be analyzed by feature extraction (a plausible but still not proven idea), subjective supplementations are not. We may conclude, therefore, either that the visual system uses a single technique to represent all information, in which case feature analysis is not a characteristic of that technique; or that the visual system has at least

---

[1] Frisby (1969) has since modified his assertions because he found a small effect of orientation on the breakdown of apparent motion after extended observation. He was not able to demonstrate such an effect on the perception of motion itself, however. I thank Neville Moray for telling me of Frisby's work.

two distinguishably different ways of representing pictorial information.

A distinction must be drawn, moreover, between the mechanisms or operations thought to be active in constructing a perception, two of which have been described, and the perceptual experience itself. Pictorially equivalent variations in apparent brightness can be created by varying the intensity of the target or by varying the intensity of the inducing flashes; visual movements may look alike in many situations, but their similarity of appearance does not testify to a similarity in construction. Many other examples can be cited to demonstrate the difference between construction or routes through the nervous system, and product or perceptual experience. Perception therefore is more like mathematics than like chemistry. A particular numerical value can be obtained by a number of different operations, but the value itself only rarely testifies to the procedure generating it. Perceptual experiences, similarly, only rarely by themselves reveal the source of their information or the course of their construction.

Although the visual system does not distinguish pictorially between environmentally-supplied and self-supplied information, the manner in which it processes the two kinds of information is notably different. Self-supplied information is not processed by the kind of feature analysis that can be described for environmentally-supplied information. It would be a mistake to think that self-supplied information is applied whimsically, however, or is not rule-governed. This will be clear from a summary of some of the characteristics discovered in the preceding investigations.

Some Characteristics of Apparent Motion

I shall summarize ten findings from the investigation. The first is that the visual system resolves or rationalizes the disparity between two properly-timed flashes. Hence the impletions or filling-in of apparent motion are not simply redundant copies of the applied stimulation; they are, rather, constructive resolutions of their difference. Second, in making a transformation, the visual object does not go through zero or shrink to a dot; it changes in the course of flight making whatever transitions are needed to resolve the disparity. In these transitions the

visual system operates in parallel, making all of the necessary adjustments at the same time. Moreover, the parallelism of operation is not
restricted to a single figure. If more than one figure needs to be transformed to rationalize a display, all of the transformations on all of the
figures are made at the same time. (Indeed, Kolers and Pomerantz
found that two transformations, size and orientation, are accomplished
as readily as either transformation alone, and no better than the better
of the single transformations. Thus the multiple transformations are
not made serially, nor do they interact synergistically so that their
joint occurrence is accomplished better than each of the single occur-
rences.)

Third, the transformations usually, but not necessarily, follow a
principle of minimal change in resolving a disparity. The illusory object
does not dart about or change speed or direction at will; it moves only
between the termini established by the physical displays, and in a manner that rationalizes their difference in relatively simple ways. Transformation R of Fig. 6.1 is an exception to the principle of minimal
change.

Fourth, the doctrine of receptive fields cannot fully explain the results.
Whether a figure is seen to undergo plastic deformation, depth, or only
blink on and off depends more upon its connection to other figures
than upon its location in a receptive field. Hence, some aspects of the
visual data affect the operations applied to them. This finding by itself
shows that a conception of pattern analysis as due to separable notions
of program and data in the visual system is not correct. Program and
data interact in visual perception.

Fifth, the visual system can create tridimensionality between two
figures when a third stimulus appears in the path of apparent motion,
when regions of excitation are within close enough range, when two
flashes are properly asynchronous, and when two contours are sufficiently strongly marked for their disparity. Compared with plastic deformation, the construction of depth takes appreciably more time.

Sixth, the figural resolutions of apparent motion are not feature-
analytic. The visual system seems to analyze physically-presented flashes
for their features, but impletions are synthetic, not analytic. In these
figural syntheses, moreover, the visual system retains a perceptual
location and distorts figures to accommodate the stimulated sites; it

does not preserve a figure. Hence, motion seems to dominate the perception and figure is dependent upon motion. The proper definition of "figure," however, is not yet known. It cannot be identified with semantic features of a display, nor with contour, nor even with spatial density. The idea that contour is fundamental or the basic unit of perception is shown to be wrong, at least for the illusion of motion.

Seventh, pictorial supplementations are sometimes susceptible to linguistic influences. People talented or skilled in the mental manipulation of pictorial representations, such as topologists or geometers, can sometimes describe resolutions of a figural disparity so that people who are less skilled pictorially can also see them; but this does not always happen. The dictionary of pictorial descriptions and the dictionary of linguistic descriptions are not identical. Indeed, we can see many things that we cannot fully describe (a landscape, for example) and we have names for things that we cannot see (devils, angels, and unicorns; plans, democracy, and intelligence).

Eighth, volition, attitude, and expectation can affect in a small way the clarity, distinctiveness, and even the maintenance of a figural construction, but only within the bounds established by the conditions of stimulation. They cannot by themselves create or determine the course of the seen movement.

Ninth, the figural aspects of the construction are postulated to occur in real time. The proposal was made that repeated presentations of the same stimulus figures allows the visual system to synthesize the pair in real time rather than delayed time. This is one way that expectation exerts an influence on the perception.

Tenth, semantic attributes of the stimuli play little if any role in the temporal and spatial limits of apparent motion. Familiarity and interpretability exert their role not on the biophysical parameters of timing, but on the clarity and convincingness of the illusory object. The role of semantic attributes and skills in their manipulation are characteristics of the synthetic supplementations, not of feature analysis.

This list is illustrative rather than exhaustive. Its main point is to demonstrate that the figural transformations of apparent motion are rule-governed, but the rules are not feature-analytic. The evidence of non-analyticity is strong enough to warrant saying either that the visual system possesses two distinctively different modes for representing information pictorially, or that feature analysis is not the means used for representing veridical stimuli.

The "picture" that we have in mind when we see has long been thought by many to have an epistemological priority in our lives. It has been thought that images or pictorial representations are the means by which physical reality is analyzed on-line. We may assert to the contrary that pictorial experience is a late product in the processing of visual information; it follows by a considerable fraction of a second the application of physical energy to the sensory transducers. Moreover, pictorial experiences ("images") can interfere with and be affected by the processing of other visual inputs (Brooks, 1967; Segal and Fusella, 1970). Pictorial experiences and images are not the means of analysis of the world but the results of that analysis, a tent the mind throws over its own operations. We see the tent, but can only infer its supports.

Features that we note in objects are the perceptual products of figural construction, but are not the primary elements of figural analysis. People perceive and report the features of objects that they have learned to pay attention to, so that most features are not obligatory or intrinsic aspects of stimuli. Perceived features, especially when they are perceptually clear, are the results of scrutiny and conjecture; they are the output of the visual system's operations as much as perceptions themselves are. The visual system having constructed a perception, its features may be used at a later time for ready identification of objects, or be used as input for other operations. These are the feed-back loops schematized in Fig. 7.6. The eyes then become servants of the person, directed here and there to seek the clues out of which pictorial experience is created, and are no longer only the passive recipients of flashes of light. The clues sought depend upon what one has learned to see, and what is seen depends upon skills applied in stocking and utilizing the pictorial dictionary. Memory and knowledge are inextricably confounded with perceiving. Lists of "distinctive features" at the graphemic level can never tell us how the visual system builds its pictorial representation of the physical world, but they may sometimes tell us what we can find in the finished product. The theory of feature analysis characteristically uses the products of perception to identify the processes of perception, and thereby sometimes limits its utility.

The impletions of apparent motion make it clear that although the visual apparatus may select from an array features to which it responds, the features themselves do not create the visual experience. Rather, that experience is generated from within, by means of supplementative mechanisms whose rules are accommodative and rationalizing rather than analytical. Historically, one group of scientists has analyzed the stimulus conditions associated with perceptions; the emphasis in their work has been on the inward flow of information. Another group of scientists has analyzed the observer; the emphasis in their work has been on the selective, supplementative, outward flow of information. Our analysis of apparent motion reveals the two-way character of information-flow, whereby the observer supplements out of his own stock information accepted from the environment, using different mechanisms for the two procedures. In this we come full circle on the problem of perception, which is to show how skills, scrutiny, and experience sensitize one to features of an infinitely complex stimulus, which are themselves the products of the visual machinery's activity. It is, presumably, the interplay of analytic and synthetic operations that enables one to learn how to see new things, and that accounts for the infinite richness of visual experience.

AFTERWORD

A NUMBER of issues are left unsettled in the foregoing pages, some empirical and others theoretical. Specification of the minimal conditions for a perception of depth, the role of color in figure perception, the meaning of the word "contour", and the characteristics of contour that permit such extraordinary plasticity in perceptual constructions are some of the empirical issues. Three other topics of a more theoretical nature bear brief comment.

I have discussed primarily the construction of visual perceptions, and have also argued, albeit briefly, that construction and decay probably follow substantially different courses. (By "decay" I mean the alterations of figure perception, both the effects and the aftereffects, following on prolonged observation.) The greatest part of our perceptual experience has not been discussed, however; that is perceptual maintenance.

Our eyes are in continual movement. Even when we look at a single object, however fixedly, the eyes move about it, saccadically, approximately three times per second. It seems to me implausible that the visual apparatus constructs a pictorial representation anew with each movement of the eyes, three times per second. It is implausible if for no other reason than that this would be an extraordinarily inefficient and wasteful procedure to go through, especially when the greatest part of what we look at remains fairly constant for seconds at a time. The procedure would be doubly wasteful in an organism that contains a visual memory. MacKay (1962) has commented on the implausibility of novel constructions with each movement of the eyes; his argument is that the visual system does not detect continuity but assumes continuity and detects only change. In other words, the visual system as a whole operates in a manner somewhat analogous to the way its peripheral receptors operate, responding most vigorously to onsets and offsets of stimulation and only weakly to continuous stimulation.

If it does operates in this fashion, then the visual system must supply

199

rom within itself the nonvarying aspects of what is seen; that is, it samples the environment, and supplements the information obtained in sampling with information from its own stores. MacKay's proposals regarding the mechanisms involved are tied tightly to eye movements, and it is just in this respect that the "cancellation", "suppression", and other theories related to the registration of motions of the eye become of special interest to students of perception. For the question concerns the means by which the visual apparatus deals with the asymmetry of motion effects produced by the same relative displacement of optical image and retina, the asymmetry described earlier produced in the one case by a stationary eye and moving object and in the other by a moving eye and stationary object. It is possible that MacKay's proposals are tied too tightly to analytic information related to eye movements themselves. The rules governing maintenance may have a more richly semantic base than eye movements would provide (Kolers, in press). Hence while we do not yet know how to account for figural maintenance, or even what its operative mechanisms are, the phenomenon of maintenance is clearly of enormous theoretical (and practical) importance: it probably accounts for the greatest part of our seeing. Consistent with what has been said in earlier chapters, perceptual maintenance seems to me to be more a phenomenon due to synthetic operations than to analytical ones. The mechanisms governing the synthetic supplementations I have described seem to be ideally suited to cope with problems of figural maintenance and perceptual continuity.

A second issue concerns the interpretation of the two kinds of operations that have been described, one for location (movement) and the other for figure. These of course are not the only operations the visual system carries out that can be characterized with analytical and synthetic routines. Phenomena of color vision seem ideally suited to a similar treatment, the feature–analytic component due to the trichromaticity of the retinal receptors (Gouras, 1970), and the pictorial experience due to the modulation of their influences (Hurvich and Jameson, 1957; Jameson and Hurvich 1964; Land and McCann, 1971, for example). Pursuing Goodman's suggestion to study color changes in apparent motion, in order to learn what kind of color space these rationalizations of disparity represent, might indeed give us an insight into the operations of this perplexing and perturbed area of visual functions. Chromatic supplementations might be found to follow different rules from those governing veridical color perception, in the same way that figural supplementations differ from feature analysis.

We can indeed then speak of three operations by means of which visual inputs are analyzed (not, however, "three visual systems"); in respect to their location, their shape, and their color. I do not know at the present time what other aspects of objects might be described in a similar way, but finding that set of features would probably provide us with the proper set of "distinctive features" of vision. This is the third topic that bears comment.

In linguistics, Jakobson's theory of distinctive features (Jakobson and Halle, 1956; Keyser and Halle, 1968; Chomsky and Halle, 1968) describes the sounds of language in terms of clusters of binary attributes. Some of these attributes are the place of articulation (the front or the back of the vocal tract, for example), the mode of articulation (open, continuing, and the like), and the energy level used in their production. Linguists have not yet agreed on the precise characterization of the set, nor even on the minimal number of features required for the complete specification of the sounds of language. Various estimates place the number at approximately a dozen, but the number itself is not terribly important. What is important is the idea of distinctive features, a set of binary attributes that reduce the data—the sounds of all the world's spoken languages—to relatively simply diagrams or tables. It is no wonder that so powerful a device has been examined avariciously by some psychologists but, alas, to little avail.

One may clip the sounds of a language electronically, one may speak a language with a different accent, through a handkerchief or with pebbles in his mouth. The features describing the sounds made vary somewhat, of course, but remarkably little. The sound /p/ remains plosive and unvoiced throughout, and the sound /n/ remains nasal. In other words, distinctive features of sounds represent certain invariants of production; and many of these physical invariants (although not all) play a role in the perception of the sounds: some distinctive features are possessed of what the linguists call "psychological reality", that is, they are important components of perception (Chomsky and Halle, 1968). What is the case for visual perception?

Visual objects do not have a similar invariance at the graphemic level.

Some binary attributes may be used to characterize them: whether the object is moving or stationary, colored or achromatic, shaped or amorphous, and the like, as suggested above. But whether an object has straight lines or curves, right-angled or left-angled diagonals, a greater or lesser perimeter-to-area ratio, and the like cannot be said to be its distinctive features, in the sense of physical invariants of the object. E. J. Gibson (1969), for example, tries to characterize the letters of the roman alphabet in terms of distinctive features, by showing that some letters have intersections, others have straight lines, and still others have curves. Such an enterprise misinterprets radically the notion of distinctive features, for there are no invariants in letters. The enormous range of type faces available for print reveal radical departures from consistency in letter shapes without a corresponding loss of identification. In one study, in fact, Kolers and Perkins (1969), commenting on the notion of distinctive features of letters, suggested that no rule can be formulated that completely specifies how a letter may be drawn; indeed, the recognition of many letters, especially but not exclusively in cursive script, requires the use of context in a way that violates altogether the notion of a completely specifiable set of features (Eden, 1968). Again the illustration of Fig. 4.2 reveals that remarkably little is available at the level of graphemic definition that one could use to characterize chairs in terms of distinctive features. The concept of distinctive features is misused by students of visual perception.

What is meant actually is not that objects have distinctive features—physical invariants—but that they have distinguishing features, bits and pieces that we have come to learn belong to or are part of a particular object. The difference between distinctive and distinguishing features is the difference between physical invariants (with a corresponding perceptual invariant) and learned properties or attributes. The visual system may indeed be possessed of a response to distinctive features; but if so, such features will have to be defined at the abstract level suggested, movement or stasis, colored or not, and the like; graphemic differences cannot do the job. The implication of this is that the visual system's response is at a level sufficiently abstract in respect of distinctive features that it allows for only the broadest or most general kinds of classification as an intrinsic property of nervous functioning; whereas the details that we use to tell objects apart, or even to tell what objects are, are based on specific entries in our perceptual dictionaries, entries put there of course by contact or commerce with the environment. We learn how to tell objects apart, not by virtue of intrinsic responses to their details, but by virtue of scrutiny, examination, attention, memory, and skills in identification. Self-supplied supplementations seem to be present in these activities from the earliest stages of encoding on; the resulting variability of our perceptual experiences would seem to be inconsistent with predictions based on the physicalistic notion of distinctive features.

REFERENCES

AARONS, L. (1964) Visual apparent movement research: Review, 1935–1955, and bibliography, 1955–1963. *Perceptual and Motor Skills*, Monograph Supplement 2-V18, **18**, 239–274.

ALLPORT, D. A. (1968) Phenomenal simultaneity and the perceptual moment hypothesis. *British Journal of Psychology*, **59**, 395–406.

ALLPORT, F. H. (1955) *Theories of Perception and the Concept of Structure*. New York: Wiley.

ALPERN, M. (1952) Metacontrast: Historical Introduction. *American Journal of Optometry and Archives of American Academy of Optometry*, **29**, 631–646.

ALPERN, M. (1953) Metacontrast. *Journal of the Optical Society of America*, **43**, 648–657.

ALPERN, M. (1954) The effect of luminance of the contrast-inducing flashes on the spatial range of metacontrast. *American Journal of Optometry*, **31**, 363–369.

AMES, A., JR. (1955) *An Interpretative Manual: the Nature of Our Perceptions, Prehensions, and Behavior*. Princeton: Princeton University Press.

ANSBACHER, H. L. (1938) Further investigation of the Harold C. Brown shrinkage phenomenon; a new approach to the study of the perception of movement. *Psychological Bulletin*, **35**, 701.

ANSBACHER, H. L. (1944) Distortion in the perception of real movement. *Journal of Experimental Psychology*, **34**, 1–23.

ARDEN, G. B. and WEALE, R. A. (1954) Variations of the latent period of vision. *Proceedings of the Royal Society (London)*, **142B**, 258–266.

AUBERT, H. (1886) Die Bewegungsempfindung. *Archiv für die gesammte Physiologie*, **39**, 347–370.

AXELROD, S. and THOMPSON, L. (1962) On visual changes of reversible figures and auditory changes in meaning. *American Journal of Psychology*, **75**, 673–674.

BARLOW, H. B. and HILL, R. M. (1963) Selective sensitivity to direction of movement in ganglion cells of the rabbit retina. *Science*, **139**, 412–414.

BARTLETT, F. C. (1932) *Remembering*. London: Cambridge University Press.

BARTLEY, S. H. (1941) *Vision, a Study of Its Basis*. New York: D. Van Nostrand.

BILL, J. C. and TEFT, L. W. (1969) Space-time relations: Effects of time on perceived visual extent. *Journal of Experimental Psychology*, **81**, 196–199.

BLAKEMORE, C. and CAMPBELL, F. W. (1969) On the existence of neurones in the human visual system selectively sensitive to the orientation and size of retinal images. *Journal of Physiology*, **203**, 237–260.

BORING, E. G. (1942) *Sensation and Perception in the History of Experimental Psychology*. New York: Appleton-Century-Crofts.

BOUMAN, M. A. and VAN DEN BRINK, G. (1953) Absolute thresholds for moving point sources. *Journal of the Optical Society of America*, **43** 895–898.

BOYNTON, R. M. (1958) On-responses in the human visual system as inferred from psychophysical studies of rapid-adaptation. *A.M.A. Archives of Ophthalmology*, Part II, **60**, 800–810.

BRAUNSTEIN, M. L. (1962) The perception of depth through motion. *Psychological Bulletin*, **59**, 422–433.

BRENNER, M. W. (1953) Continuous stimulation and apparent movement. *American Journal of Psychology*, **66**, 494–495.

BRENNER, M. W. (1957) The developmental study of apparent movement. *Quarterly Journal of Experimental Psychology*, **9**, 169–174.

BROADBENT, D. E. (1958) *Perception and Communication*. New York: Pergamon.

BROOKS, L. R. (1967) The suppression of visualization by reading. *Quarterly Journal of Experimental Psychology*, **19**, 289–299.

BROSGOLE, L. (1966) Change in phenomenal location and perception of motion. *Perceptual and Motor Skills*, **23**, 999–1001.

BROWN, J. F. (1931) The perception of visual velocity. *Psychologische Forschung*, **14**, 199–232.

BROWN, J. F. (1931) On time perception in visual movement fields. *Psychologische Forschung*, **14**, 233–248.

BROWN, J. F. (1931) The thresholds for visual movement. *Psychologische Forschung*, **14**, 249–268.

BROWN, J. F. and VOTH, A. C. (1937) The path of seen movement as a function of the vector-field. *American Journal of Psychology*, **49**, 543–563.

BROWN, R. H. (1958) Influence of stimulus luminance upon the upper speed threshold for the visual discrimination of movement. *Journal of the Optical Society of America*, **48**, 125–128.

BROWN, R. H. (1961) Visual sensitivity to differences in velocity. *Psychological Bulletin*, **58**, 89–103.

BRUNER, J. S., POSTMAN, L. and MOSTELLER, F. (1950) A note on the measurement of reversals of perspective. *Psychometrika*, **15**, 63–72.

CAMPBELL, F. W. and MAFFEI, L. (1970) Electrophysiological evidence for the existence of orientation and size detectors in the human visual system. *Journal of Physiology*, **207**, 635–652.

CHOMSKY, N. and HALLE, M. (1968) *The Sound Pattern of English*. New York: Harper & Row.

COHEN, J., HANSEL, C. E. M. and SYLVESTER, J. D. (1953) A new phenomenon in time judgment. *Nature*, **172**, 901.

COHEN, J., HANSEL, C. E. M. and SYLVESTER, J. D. (1955) Interdependence in judgments of space, time and movement. *Acta Psychologica*, **11**, 360–372.

COHEN, L. (1959) Rate of apparent change of a Necker Cube as a function of prior stimulation. *American Journal of Psychology*, **72**, 327–344.

COHEN, R. L. (1964) *Problems in Motion Perception*. Uppsala: Appelbergs.

CORNSWEET, T. N. (1956) Determination of the stimuli for involuntary drifts and saccadic eye movements. *Journal of the Optical Society of America*, **46**, 987–993.

CORWIN, T. R. and BOYNTON, R. M. (1968) Transitivity of visual judgments of simultaneity. *Journal of Experimental Psychology*, **78**, 560–568.

CRAWFORD, B. H. (1947) Visual adaptation in relation to brief conditioning stimuli. *Proceedings of the Royal Society (London)*, **134B**, 283–302.

DEATHERAGE, B. H. and BITTERMAN, M. E. (1952) The effect of satiation on stroboscopic movement. *American Journal of Psychology*, **65**, 108–109.

DeSilva, H. R. (1926) An experimental investigation of the determinants of apparent visual movement. *The American Journal of Psychology*, **37**, 469–501.

DeSilva, H. R. (1928) Kinematographic movement of parallel lines. *The Journal of General Psychology*, **1**, 550–577.

DeSilva, H. R. (1929) An analysis of the visual perception of movement. *British Journal of Psychology*, **19**, 268–305.

Dimmick, F. L. and Sanders, R. W. (1929) Some conditions of the perception of visible movement. *American Journal of Psychology*, **41**, 607–616.

Djang, S. (1937) The role of past experience in the visual apprehension of masked forms. *Journal of Experimental Psychology*, **20**, 29–59.

Dodge, R. (1907) An experimental study of visual fixation. *Psychological Review Monograph Supplement*, **8** (Whole No. 35).

Dodwell, P. C. (1970) *Visual Pattern Recognition*. New York: Holt, Rinehart & Winston.

Duncker, K. (1929) Über induzierte Bewegung. *Psychologische Forschung*, **12**, 180–259.

Eden, M. (1968) Handwriting generation and recognition. In P. A. Kolers and M. Eden (eds.), *Recognizing Patterns*. Cambridge, Mass.: M.I.T. Press.

Efron, R. (1957) Stereoscopic vision: I. Effect of binocular temporal summation. *British Journal of Ophthalmology*, **61**, 709–730.

Efron, R. (1967) The duration of the present. *Annals of the New York Academy of Sciences*, **138** (2), 713–729.

Enright, J. T. (1970) Distortions of apparent velocity: A new optical illusion. *Science*, **168**, 464–467.

Ewert, P. H. (1930) The perception of visible movement. *Psychological Bulletin*, **27**, 318–328.

Exner, S. (1875) Ueber das Sehen von Bewegungen und die Theorie des zusammengesetzen Auges. *Sitzungsberichte Akademie Wissenschaft Wien*, **72**, 156–190.

Fadiga, E. and Pupilli, G. C. (1964) Teleceptive components of the cerebellar function. *Physiological Reviews*, **44**, 432–486.

Fehrer, E. and Raab, D. (1962) Reaction time to stimuli masked by metacontrast. *Journal of Experimental Psychology*, **63**, 143–147.

Fehrer, E. and Smith, E. (1962) Effect of luminance ratio on masking. *Perceptual and Motor Skills*, **14**, 243–253.

Fender, D. and Julesz, B. (1967) Extension of Panum's fusional area in binocularly stabilized vision. *Journal of the Optical Society of America*, **57**, 819–830.

Fischer, G. J. (1956) Factors affecting estimation of depth with variations of the stereokinetic effect. *American Journal of Psychology*, **69**, 252–257.

Fisichelli, V. R. (1946) Effect of rotational axis and dimensional variations on the reversals of apparent movement in Lissajous figures. *American Journal of Psychology*, **59**, 669–675.

Fraisse, P. (1963) *The Psychology of Time*. New York: Harper & Row.

Freedman, S. J. (ed.) (1968) *The Neuropsychology of Spatially Oriented Behavior*. Homewood, Illinois: Dorsey.

Frisby, J. P. (1968) Some relationships between thresholds for apparent movement and orientation of stimuli. *Bulletin of the British Psychological Society*, **21**, 31 (abstract).

Frisby, J. P. (1969) The effect of stimulus orientation on the phi phenomenon. Ph.D. thesis, University of Sheffield.

FRY, G. A. and BARTLEY, S. H. (1935) The effect of one border in the visual field upon the threshold of another. *American Journal of Physiology*, 112, 414–421.

GALLI, P. A. (1932) Ueber mittelst verschiedener Sinnesreize erweckte Wahrnehmung von Scheinbewegungen. *Archiv für die gesammte Psychologie*, 85, 137–180.

GAZE, R. M., KEATING, M. J., SZÉKELEY, G., and BEAZLEY, L. (1970) Binocular interaction in the formation of specific intertectal neuronal connexions. *Proceedings of the Royal Society (London)*, 175B, 107–147.

GELDREICH, E. W. (1934) A lecture room demonstration of the visual tau-effect. *American Journal of Psychology*, 46, 483.

GENGERELLI, J. A. (1948) Apparent movement in relation to homonymous and heteronymous stimulation of the cerebral hemispheres. *The Journal of Experimental Psychology*, 38, 592–599.

GIBSON, E. J. (1969) *Principles of Perceptual Learning and Development*. New York: Appleton-Century-Crofts.

GIBSON, J. J. (1954) The visual perception of objective motion and subjective movement. *Psychological Review*, 61, 304–314.

GIBSON, J. J. (1968) What gives rise to the perception of motion? *Psychological Review*, 75, 335–346.

GILINSKY, A. S. and DOHERTY, R. S. (1969) Interocular transfer of orientational effects. *Science*, 164, 454–455.

GOLDSTEIN, A. G. (1957) Judgments of visual velocity as a function of length of observation time. *Journal of Experimental Psychology*, 54, 457–461.

GOLDSTEIN, J. and WIENER, C. (1963) On some relations between the perception of depth and of movement. *The Journal of Psychology*, 55, 3–23.

GOLDSTEIN, J. and WIENER, C. (1969) Visual movement and the bending of phenomenal space. *Journal of General Psychology*, 80, 3–46.

GOMBRICH, E. H. (1961) *Art and Illusion*. Second edition, New York: Pantheon.

GOODMAN, N. (1968) *Languages of Art: an Approach to a Theory of Symbols*. Indianapolis and New York: Bobbs-Merrill.

GOODMAN, N. (1970) Seven strictures on similarity. In L. Foster and J. W. Swanson (eds.), *Experience and Theory*. Amherst: University of Massachusetts Press.

GOTTSCHALDT, K. (1926) Ueber den Einfluss der Erfahrung auf die Wahrnehmung von Figuren. *Psychologische Forschung*, 8, 261–317.

GOURAS, P. (1970) Trichromatic mechanisms in single cortical neurons. *Science*, 168, 489–492.

GRAHAM, C. H. (1951) Visual perception. In S. S. Stevens (ed.) *Handbook of Experimental Psychology*. New York: Wiley.

GREGORY, R. L. (1966) *Eye and Brain*. New York: McGraw-Hill.

GRÜNBAUM, A. (1967) *Modern Science and Zeno's Paradoxes*. Middletown, Conn.: Wesleyan University Press.

GRUNDFEST, H. (1967) Synaptic and ephaptic transmission. In G. C. Quarton, T. Melnechuk, and F. O. Schmitt (eds.), *The Neurosciences*. New York: Rockefeller University Press.

GRÜSSER-CORNEHLS, U., GRÜSSER, O.-J. and BULLOCK, T. H. (1963) Unit responses in the frog's tectum to moving and non-moving visual stimuli. *Science*, 141, 820–822.

GUILFORD, J. P. and HELSON, H. (1929) Eye-movements and the phi-phenomenon. *American Journal of Psychology*, 41, 595–606.

HABER, R. N. and HERSHENSON, M. (1965) The effects of repeated brief exposures on the growth of a percept. *Journal of Experimental Psychology*, **69**, 40–46.

HARRIS, C. S. (1965) Perceptual adaptation to inverted, reversed, and displaced vision. *Psychological Review*, **72**, 419–444.

HARTER, M. R. (1967) Excitability cycles and cortical scanning: A review of two hypotheses of central intermittency in perception. *Psychological Bulletin*, **68**, 47–58.

HARTLINE, H. K. (1934) Intensity and duration in the excitation of single photoreceptor units. *Journal of Cellular and Comparative Physiology*, **5**, 229–247.

HARTMANN, G. W. (1935) *Gestalt Psychology*, New York: Ronald Press.

HEBB, D. O. (1949) *The Organization of Behavior*. New York: Wiley.

HEIDBREDER, E. (1933) *Seven Psychologies*. New York: Appleton-Century-Crofts.

HELSON, H. and KING, S. M. (1931) The tau effect: An example of psychological relativity. *Journal of Experimental Psychology*, **14**, 202–217.

HIGGINSON, G. D. (1926) Apparent visual movement and the Gestalt. *Journal of Experimental Psychology*, **9**, 228–252.

HIGGINSON, G. D. (1926) The place of ocular movements in stroboscopic perception. *American Journal of Psychology*, **37**, 408–413.

HIRSH, I. J. and SHERRICK, C. E., Jr. (1961) Perceived order in different sense modalities. *Journal of Experimental Psychology*, **62**, 423–432.

HOVLAND, C. I. (1935) Apparent movement. *Psychological Bulletin*, **32**, 755–778.

HOWARD, I. P. and TEMPLETON, W. B. (1966) *Human Spatial Orientation*. New York: Wiley.

HUBEL, D. H. and WIESEL, T. N. (1962) Receptive fields, binocular interaction, and functional architecture in the cat's visual cortex. *Journal of Physiology*, **10**, 106–154.

HUBEL, D. H. and WIESEL, T. N. (1965) Receptive fields and functional architecture in two nonstriate visual areas (18 and 19) of the cat. *Journal of Neurophysiology*, **28**, 229–289.

HUEY, E. B. (1968) *The Psychology and Pedagogy of Reading*. New York: Macmillan. Reissued Cambridge, Massachusetts: M.I.T. Press, 1968.

HUMPHREY, G. and SPRINGBETT, B. M. (1946) The after-image of the phi-phenomenon. *Canadian Psychological Association Bulletin*, **6**, 3–6.

HURVICH, L. M. and JAMESON, D. (1957) An opponent-process theory of color vision. *Psychological Review*, **64**, 384–404.

INGLE, D. (1967) Two visual mechanisms underlying the behavior of fish. *Psychologische Forschung*, **31**, 44–51.

IVINS, W. M., JR. (1968) *Prints and Visual Communication*. Cambridge, Massachusetts: M.I.T. Press.

JACOBS, G. H. (1969) Receptive fields in visual systems. *Brain Research*, **14**, 553–573.

JAKOBSON, R. and HALLE, M. (1956) *Fundamentals of Language*. The Hague: Mouton.

JAMESON, D. and HURVICH, L. M. (1964) Theory of brightness and color contrast in human vision. *Vision Research*, **4**, 135–154.

JEEVES, M. A. and BRUNER, J. S. (1956) Directional information and apparent movement. *The Quarterly Journal of Experimental Psychology*, **8**, 107–113

JOHANSSON, G. (1950) *Configurations in Event Perception*. Uppsala, Sweden: Almqvist & Wiksell.

JONES, E. E. and BRUNER, J. S. (1954) Expectancy in apparent visual movement. *British Journal of Psychology*, **45**, 157–165.

JULESZ, B. (1960) Binocular depth perception of computer generated patterns. *Bell System Technical Journal*, **39**, 1125–1161.

JULESZ, B. (1971) *Foundations of Cyclopean Perception*. Chicago: University of Chicago Press.

KAHNEMAN, D. (1967) An onset-onset law for one case of apparent motion and meta-contrast. *Perception and Psychophysics*, **2**, 577–584.

KAHNEMAN, D. (1968) Method, findings, and theory in studies of visual masking. *Psychological Bulletin*, **70**, 404–425.

KALISH, H. I. (1969) Stimulus generalization. In M. H. Marx (ed.), *Learning: Processes*. New York: Macmillan.

KELLY, E. L. (1935) The effect of previous experience and suggestion on the perception of apparent movement. *Psychological Bulletin*, **32**, 569–570.

KEYSER, S. J. and HALLE, M. (1968) What we do when we speak. In P. A. Kolers and M. Eden (eds.), *Recognizing Patterns*. Cambridge, Massachusetts: M.I.T. Press.

KINCHLA, R. A. and ALLAN, L. G. (1969) A theory of visual movement perception. *Psychological Review*, **76**, 537–558.

KLEMMER, E. T. (1957) Rhythmic disturbances in a simple visual-motor task. *The American Journal of Psychology*, **70**, 56–63.

KLEMMER, E. T. (1967) Sequences of responses to signals encoded in time only. *Acta Psychologica*, **27**, 197–203.

KOFFKA, K. (1931) Die Wahrnehmung von Bewegung. In A. Bethe *et al.* (eds.), *Handbuch der normalen und pathologische Physiologie*, vol. 12, part 2, pp. 1166–1214. Berlin: Springer.

KOFFKA, K. (1935) *Principles of Gestalt Psychology*. New York: Harcourt, Brace.

KÖHLER, W. (1923) Zur Theorie der stroboskopischen Bewegung. *Psychologische Forschung*, **3**, 397–406.

KÖHLER, W. (1947) *Gestalt Psychology*. New York: Liveright.

KÖHLER, W. and WALLACH, H. (1944) Figural after-effects. *Proceedings of the American Philosophical Society*, **88**, 269–357.

KÖHLER, W., WALLACH, H. and CARTWRIGHT, D. (1942) Two theories of visual speed. *The Journal of General Psychology*, **27**, 93–109.

KOLERS, P. A. (1962) Intensity and contour effects in visual masking. *Vision Research*, **2**, 277–294.

KOLERS, P. A. (1963) Some differences between real and apparent visual movement. *Vision Research*, **3**, 191–206.

KOLERS, P. A. (1964) The illusion of movement. *Scientific American*, **211** (4), 98–106.

KOLERS, P. A. (1964) Apparent movement of a Necker Cube. *American Journal of Psychology*, **77**, 220–230.

KOLERS, P. A. (1966) An illusion that dissociates motion, object, and meaning. *Quarterly Progress Report No. 82, Research Laboratory of Electronics*, M.I.T. July 15, 1966, pp. 221–223.

KOLERS, P. A. (1966) Reading and talking bilingually. *American Journal of Psychology*, **79**, 357–376.

KOLERS, P. A. (1968) Some psychological aspects of pattern recognition. In P. A. Kolers and M. Eden (eds.), *Recognizing Patterns*. Cambridge, Massachusetts: M.I.T. Press.

KOLERS, P. A. (1969) Voluntary attention switching between foresight and hindsight. *Quarterly Progress Report No. 92, Research Laboratory of Electronics*, M.I.T. January 15, 1969, pp. 381–385.

KOLERS, P. A. (1970) The role of shape and geometry in picture recognition. In B. S. Lipkin and A. Rosenfeld (eds.), *Psychopictorics and Picture Processing.* New York: Academic Press.

KOLERS, P. A. (In press) Reading pictures: Some cognitive aspects of visual perception. In T. Huang and O. Tretiak (eds.), *Picture Bandwidth Compression.*

KOLERS, P. A. and PERKINS, D. N. (1969) Orientation of letters and their speed of recognition. *Perception and Psychophysics,* **5,** 275–280.

KOLERS, P. A. and POMERANTZ, J. R. (1971) Figural change in apparent motion. *Journal of Experimental Psychology,* **87,** 99–108.

KOLERS, P. A. and ROSNER, B. S. (1960) On visual masking (metacontrast): dichoptic observation. *American Journal of Psychology,* **73,** 2–21.

KOLERS, P. A. and TOUCHSTONE, G. E. (1965) Variations of perceived distance with apparent motion. *Quarterly Progress Report No. 79, Research Laboratory of Electronics,* Massachusetts Institute of Technology, October 15, 1965, 207–211.

KOLERS, P. A. and ZINK, D. L. (1962) Some aspects of problem solving: sequential analysis of the detection of embedded patterns. AMRL-TDR-62-148, Wright-Patterson Air Force Base, Ohio.

KORTE, A. (1915) Kinematoskopische Untersuchungen. *Zeitschrift für Psychologie,* **72,** 194–296.

KUFFLER, S. W. (1953) Discharge patterns and functional organization of mammalian retina. *Journal of Neurophysiology,* **16,** 37–68.

KÜLPE, O. (1893) *Grundriss der Psychologie.* Leipzig: Engelmann. Translated by E. B. Titchener, *Outlines of Psychology.* New York: Macmillan, 1895.

LAND, E. H. and McCANN, J. J. (1971) Lightness and retinex theory. *Journal of the Optical Society of America,* **61,** 1–11.

LASHLEY, K. S. (1951) The problem of serial order in behavior. In L. A. Jeffress (ed.), *Cerebral Mechanisms in Behavior.* New York: Wiley.

LEGRAND, Y. (1967) *Form and Space Vision.* Bloomington: Indiana University Press.

LEIBOWITZ, H. W. (1955a) Effect of reference lines on the discrimination of movement. *Journal of the Optical Society of America,* **45,** 829–830.

LEIBOWITZ, H. W. (1955b) The relation between the rate threshold for the perception of movement and luminance for various durations of exposure. *Journal of Experimental Psychology,* **49,** 209–214.

LETTVIN, J. Y., MATURANA, H. R., McCULLOCH, W. S. and PITTS, W. H. (1959) What the frog's eye tells the frog's brain. *Proceedings of the Institute of Radio Engineers, New York,* **47,** 1940–1951.

LEWIN, K. (1931) The conflict between Aristotelian and Galilean modes of thought in contemporary psychology. *Journal of General Psychology,* **5,** 141–177.

LICHTENSTEIN, M., WHITE, C. T., SIEGFRIED, J. B. and HARTER, M. R. (1963) Apparent rate of flicker of various retinal loci and number of perceived flashes per unit time: A paradox. *Perceptual and Motor Skills,* **71,** 523–536.

LONDON, I. D. (1954) Research on sensory interaction in the Soviet Union. *Psychological Bulletin,* **51,** 531–568.

LURIA, S. M. (1965) Effects of continuously and discontinuously moving stimuli on the luminance threshold of a stationary stimulus. *Journal of the Optical Society of America,* **55,** 418–425.

LURIA, S. M. and KOLERS, P. A. (1962) Interaction of moving and stationary visual stimuli. *Journal of the Optical Society of America,* **52,** 1320 (abstract).

MacKay, D. M. (1962) Theoretical models of space perception. In C. A. Muses (ed.), *Aspects of the Theory of Artificial Intelligence*. New York: Plenum Press.

MacKay, D. M. (1970) Elevation of visual threshold by displacement of retinal image. *Nature*, 225, 90–92.

MacKay, D. M. (1970) Interocular transfer of suppressive effects of retinal image displacement. *Nature*, 225, 872–873.

Mandriota, F. J., Mintz, D. E. and Notterman, J. M. (1962) Visual velocity discrimination: Effects of spatial and temporal cues. *Science*, 138, 437–438.

Marks, E. S. (1933) The effect of boundaries on space estimation. *Journal of General Psychology*, 8, 467–472.

Marshall, A. J. and Stanley, G. (1964) The apparent length of light and dark arcs seen peripherally in rotary motion. *Australian Journal of Psychology*, 16, 120–128.

Mashhour, M. (1964) *Psychophysical Relations in the Perception of Velocity*. Stockholm: Almqvist & Wiksell.

Mashhour, M. (1969) A study of motion preferences. *Scandinavian Journal of Psychology*, 10, 299–305.

Mates, B. and Graham, C. H. (1970) Effect of rectangle length on velocity thresholds for real movement. *Proceedings of the National Academy of Sciences*, 65, 516–520.

Matin, L. and MacKinnon, G. E. (1964) Autokinetic movement: selective manipulation of directional components by image stabilization. *Science*, 143, 147–148.

Matin, L., Matin, E. and Pearce, D. G. (1969) Visual perception of direction when voluntary saccades occur: I. Relation of visual direction of a fixation target extinguished before a saccade to a flash presented during the saccade. *Perception and Psychophysics*, 5, 65–80.

Matin, L., Matin, E. and Pola, J. (1970) Visual perception of direction when voluntary saccades occur: II. Relation of visual direction of a fixation target extinguished before a saccade to a subsequent test flash presented before the saccade. *Perception and Psychophysics*, 8, 9–14.

Mayzner, M. S. and Tresselt, M. E. (1970) Visual information processing with sequential inputs: a general model for sequential blanking, displacement, and overprinting phenomena. *Annals of the New York Academy of Sciences*, 169, 599–618.

McConnell, R. F. (1927) Visual movement under simultaneous excitations with initial and terminal overlap. *Journal of Experimental Psychology*, 10, 227–246.

Miles, W. R. (1931) Movement interpretations of the silhouette of a revolving fan. *American Journal of Psychology*, 43, 392–405.

Miller, J. W. and Ludvigh, E. (1961) The perception of movement persistence in the Ganzfeld. *Journal of the Optical Society of America*, 51, 57–60.

Moray, N. (1969) *Attention: Selective Processes in Vision and Hearing*. London: Hutchinson Educational.

Morinaga, S., Noguchi, K. and Yokoi, K. (1966) Direct comparison of real and apparent visual movement. *Perceptual and Motor Skills*, 22, 346.

Muijen, A. R. W. (1969) Rhythmic registration of movement, apparent movement and apparent rest. *Acta Psychologica*, 29, 134–149.

Neff, W. S. (1936) A critical investigation of the visual apprehension of movement. *American Journal of Psychology*, 48, 1–42.

Neisser, U. (1967) *Cognitive Psychology*. New York: Appleton-Century-Crofts.

NEUHAUS, W. (1930) Experimentelle Untersuchung der Scheinbewegung. *Archiv für die gesamte Psychologie*, **75**, 315–458.

OGLE, K. N. (1950) *Researches in Binocular Vision*. Philadelphia: Saunders.

OGLE, K. N. (1963) Stereoscopic depth perception and exposure delay between the images to the two eyes. *Journal of the Optical Society of America*, **53**, 1296–1304.

OLDFIELD, R. C. (1948) La perception visuelle des images du cinema, de la television et du radar. *Revue Internationale de Filmologie*, **1**, 16–32.

OLSON, R. K. and ATTNEAVE, F. (1970) What variables produce similarity grouping? *American Journal of Psychology*, **83**, 1–21.

ORBACH, J., EHRLICH, D., and HEATH, H. A. (1963) Reversibility of the Necker Cube: I. An examination of the concept of "satiation of orientation". *Perceptual and Motor Skills*, **17**, 439–458.

ORLANSKY, J. (1940) The effect of similarity and difference in form on apparent visual movement. *Archives of Psychology*, 246.

OSGOOD, C. E. (1953) *Method and Theory in Experimental Psychology*. New York: Oxford University Press.

PANTLE, A. and SEKULER, R. (1969) Contrast response of human visual mechanisms sensitive to orientation and direction of motion. *Vision Research*, **9**, 397–406.

PANUM, P. L. (1858) *Physiologische Untersuchungen über das Sehen mit zwei Augen*. Kiel.

PERKY, C. W. (1910) An experimental study of imagination. *American Journal of Psychology*, **21**, 422–452.

POLLOCK, W. T. (1953) The visibility of a target as a function of its speed of movement. *Journal of Experimental Psychology*, **45**, 449–454.

POLYAK, S. L. (1957) *The Vertebrate Visual System*. Chicago: University of Chicago Press.

POMERANTZ, J. R. (1970) Eye movements affect the perception of apparent (beta) movement. *Psychonomic Science*, **19**, 193–194.

POSNER, M. I. (1969) Abstraction and the process of recognition. In G. H. Bower and J. T. Spence (eds.), *The Psychology of Learning and Motivation*, vol. 3. New York: Academic Press.

POSNER, M. I. and MITCHELL, R. F. (1967) Chronometric analysis of classification. *Psychological Review*, **74**, 392–409.

PRITCHARD, R. M. (1961) Stabilized images on the retina. *Scientific American*, **204** (6), 72–78.

PRITCHARD, R. M., HERON, W. and HEBB, D. O. (1960) Visual perception approached through the method of stabilized images. *Canadian Journal of Psychology*, **14**, 67–77.

PRYLUCK, C. and SNOW, R. E. (1967) Toward a psycholinguistics of cinema. *AV Communications Review*, **15**, 54–75.

RAAB, D. H. (1963) Backward masking. *Psychological Bulletin*, **60**, 118–129.

RAAB, D. and FEHRER, E. (1962) Supplementary report: the effect of stimulus duration and luminance on visual reaction time. *Journal of Experimental Psychology*, **64**, 326–327.

RASKIN, L. M. (1969) Long-term memory effects in the perception of apparent movement. *Journal of Experimental Psychology*, **79**, 97–103.

RATLIFF, F. (1961) Inhibitory interaction and the detection and enhancement of contours. In W. A. Rosenblith (ed.), *Sensory Communication*. New York: M.I.T. Press and John Wiley.

RIGGS, L. A., RATLIFF, F., CORNSWEET, J. C., and CORNSWEET, T. (1953) The disappearance of steadily fixated visual test objects. *Journal of the Optical Society of America*, **43**, 495–501.

ROCK, I. and EBENHOLTZ, S. (1962) Stroboscopic movement based on change of phenomenal rather than retinal location. *The American Journal of Psychology*, **75**, 193–207.

ROCK, I., TAUBER, E. S. and HELLER, D. P. (1965) Perception of stroboscopic movement: Evidence for its innate basis. *Science*, **147**, 1050–1052.

ROSS, P. L. (1967) Accuracy of judgments of movement in depth from two-dimensional projections. *Journal of Experimental Psychology*, **75**, 217–225.

ROYCE, J. R., CARRAN, A. B., AFTANAS, M., LEHMAN, R. S. and BLUMENTHAL, A. (1966) The autokinetic phenomenon: A critical review. *Psychological Bulletin*, **65**, 243–260.

RUSSELL, B. (1938) *The Principles of Mathematics*, 2nd ed. New York: W. W. Norton.

RUSSELL, B. (1945) *A History of Western Philosophy*. New York: Simon & Schuster.

SCHILLER, P. H. (1965) Backward masking for letters. *Perceptual and Motor Skills*, **20**, 47–50.

SCHNEIDER, G. E. (1969) Two visual systems. *Science*, **163**, 895–902.

SCHOLZ, W. (1924) Experimentelle Untersuchungen über die phänomenale Grösse von Raumstrecken, die durch Sukzessiv-Darbietung zweier Reize begrenzt werden. *Psychologische Forschung*, **5**, 219–272.

SEGAL, S. J. and BARR, H. L. (1969) Effect of instructions on phi phenomenon, criterion task of "tolerance for unrealistic experiences". *Perceptual and Motor Skills*, **29**, 483–486.

SEGAL, S. J. and FUSELLA, V. (1970) Influence of imaged pictures and sounds on detection of visual and auditory signals. *Journal of Experimental Psychology*, **83**, 458–464.

SEGAL, S. J. and GORDON, P. (1969) The Perky effect revisited: Blocking of visual signals by imagery. *Perceptual and Motor Skills*, **28**, 791–797.

SEKULER, R. W. and GANZ, L. (1963) Aftereffect of seen motion with a stabilized retinal image. *Science*, **139**, 419–420.

SEKULER, R. and PANTLE, A. (1967) A model for after-effects of seen movement. *Vision Research*, **7**, 427–439.

SGRO, F. J. (1963) Beta motion thresholds. *Journal of Experimental Psychology*, **66**, 281–285.

SHAFFER, O. and WALLACH, H. (1966) Extent-of-motion thresholds under subject-relative and object-relative conditions. *Perception and Psychophysics*, **1**, 447–451.

SHAPIRO, M. B. (1954) A preliminary investigation of the effects of continuous stimulation on the perception of "apparent motion". *British Journal of Psychology*, **45**, 58–67.

SHERRICK, C. E. (1968) Bilateral apparent haptic movement. *Perception and Psychophysics*, **4**, 159–162.

SHERRICK, C. E. and ROGERS, R. (1966) Apparent haptic movement. *Perception and Psychophysics*, **1**, 175–180.

SHERRINGTON, C. S. (1897) On reciprocal action in the retina as studied by means of some rotating discs. *Journal of Physiology*, **21**, 33–54.

SHERRINGTON, C. S. (1906) *The Integrative Action of the Nervous System*. New Haven: Yale University Press.

SHIPLEY, T. (1961) *Classics in Psychology*. New York: Philosophical Library.

SHIPLEY, W. C., KENNEY, F. A. and KING, M. E. (1945) Beta apparent movement under binocular, monocular, and interocular stimulation. *American Journal of Psychology*, 58, 545–549.

SMITH, K. R. (1948) Visual apparent movement in the absence of neural interaction. *American Journal of Psychology*, 61, 73–78.

SMITH, K. U. (1941) Experiments on the neural basis of movement vision. *Journal of Experimental Psychology*, 28, 199–216.

SMITH, M. C. (1967) Theories of the psychological refractory period. *Psychological Bulletin*, 67, 202–213.

SMITH, O. W. and SHERLOCK, L. (1957) A new explanation of the velocity-transposition phenomenon. *Americal Journal of Psychology*, 70, 102–105.

SMITH, W. M. and GULICK, W. L. (1956) Visual contour and movement perception. *Science*, 124, 316–317.

SMITH, W. M. and GULICK, W. L. (1957) Dynamic contour perception, *Journal of Experimental Psychology*, 53, 145–152.

SQUIRES, P. C. (1928) Apparent movement. *Psychological Bulletin*, 25, 245–260.

SQUIRES, P. C. (1931) The influence of hue on apparent visual movement. *American Journal of Psychology*, 43, 49–64.

STANLEY, G. (1964) A study of some variables influencing the Ansbacher shrinkage effect. *Acta Psychologica*, 22, 109–118.

STANLEY, G. (1966) Apparent shrinkage of a rotating arc as a function of luminance relations between figure and surround. *Acta Psychologica*, 25, 357–364.

STANLEY, G. (1967) Apparent brightness of a rotating arc-line as a function of speed of rotation. *Acta Psychologica*, 26, 17–21.

STANLEY, G. (1970) Static visual noise and the Ansbacher effect. *Quarterly Journal of Experimental Psychology*, 22, 43–48.

STEINIG, K. (1929) Untersuchungen über die Wahrnehmung der Bewegung durch das Auge. *Zeitschrift für Psychologie*, 109, 291–336.

STERNBERG, S. (1967) Two operations in character-recognition: Some evidence from reaction-time measurements. *Perception and Psychophysics*, 2, 45–53.

STERNBERG, S. (1970) Memory-scanning: mental processes revealed by reaction-time experiments. In J. S. Antrobus (ed.), *Cognition and Affect*. Boston: Little, Brown.

STOPER, A. E. (1963) On the perception of motion. Unpublished Master's thesis, Brandeis University.

STOPER, A. E. (1967) Vision during pursuit movement: the role of oculomotor information. Unpublished doctoral dissertation, Brandeis University.

STROUD, J. M. (1956) The fine structure of psychological time. In H. Quastler (ed.), *Information Theory in Psychology*. Glencoe, Illinois: The Free Press.

SUTHERLAND, N. S. (n.d.) The methods and findings of experiments on the visual discrimination of shape by animals. *Experimental Psychology Society*, Monograph No. 1.

SUTHERLAND, N. S. (1968) Outlines of a theory of visual pattern recognition in animals and man. *Proceedings of the Royal Society (London)*, 171B, 297–317.

SWEET, A. L. (1953) Temporal discrimination by the human eye. *American Journal of Psychology*, 66, 185–198.

SYLVESTER, J. (1960) Apparent movement and the Brown–Voth experiment. *Quarterly Journal of Experimental Psychology*, 12, 231–236.

TAKALA, M. (1951) Asymmetries of the visual space. *Annales Academiae Scientiarum Fennicae*, Series B, 72, 2.

TERNUS, J. (1926) Experimentelle Untersuchungen über phänomenale Identität. *Psychologische Forschung*, **7**, 81–136. Abstracted and translated in W. D. Ellis (ed.), *A Sourcebook of Gestalt Psychology*. New York: Humanities Press, 1950.

TEUBER, H.-L. (1960) Perception. In J. Field, H. W. Magoun, and V. E. Hall (eds.), *Handbook of Physiology—Neurophysiology III*. Washington, D.C.: American Physiological Society.

TEUBER, H.-L., BATTERSBY, W. S., and BENDER, M. B. (1960) *Visual Field Defects after Penetrating Missile Wounds of the Brain*. Cambridge, Mass.: Harvard University Press.

TEUBER, H.-L. and BENDER, M. B. (1950) Perception of apparent movement across acquired scotomata in the visual field. *American Psychologist*, **5**, 271 (abstract).

THORSON, J., LANGE, G. D., and BIEDERMAN-THORSON, M. (1969) Objective measure of the dynamics of a visual movement illusion. *Science*, **164**, 1087–1088.

TITCHENER, E. B. (1902) *An Outline of Psychology*. New York: Macmillan.

TOCH, H. H. (1956) The perceptual elaboration of a stroboscopic presentation. *American Journal of Psychology*, **69**, 345–358.

TOCH, H. H. and ITTELSON, W. H. (1956) The role of past experience in apparent movement: A revaluation. *The British Journal of Psychology*, **47**, 195–207.

UTTAL, W. R. (1970) Violations of visual simultaneity. *Perception and Psychophysics*, **7**, 133–136.

VAN DER WAALS, H. G. and ROELOFS, C. O. (1930) Optische Scheinbewegung. *Zeitschrift für Psychologie und Physiologie des Zinnesorgane*, **114**, 241–288 (1930); **115**, 91–190 (1931).

VON BÉKÉSY, G. (1967) *Sensory Inhibition*. Princeton, N.J.: Princeton University Press.

VON BÉKÉSY, G. (1969) Similarities of inhibition in the different sense organs. *American Psychologist*, **24**, 707–719.

VON BÉKÉSY, G. (1969) The smallest time difference the eyes can detect with sweeping stimulation. *Proceedings of the National Academy of Sciences*, **64**, 142–147.

VON SCHILLER, P. (1933) Stroboskopische Alternativversuche. *Psychologische Forschung*, **17**, 179–214.

VON SZILY, A. (1905) Bewegungsnachbild und Bewegungskontrast. *Zeitschrift für Psychologie und Physiologie der Sinnesorgane*, **38**, 81–154.

WALLACH, H. (1959) Perception of motion. *Scientific American*, **201** (1), 56–60.

WALLACH, H. and O'CONNELL, D. N. (1953) The kinetic depth effect. *Journal of Experimental Psychology*, **45**, 205–217.

WALLACH, H., O'CONNELL, D. N., and NEISSER, U. (1953) The memory effect of visual perception of three dimensional form. *Journal of Experimental Psychology*, **45**, 360–368.

WALLACH, H., WEISZ, A., and ADAMS, P. (1956) Circles and derived figures in rotation. *American Journal of Psychology*, **69**, 48–59.

WALLS, G. L. (1963) *The Vertebrate Eye and Its Adaptive Radiation*. New York: Hafner.

WERNER, H. (1935) Studies on contour: I. Qualitative analyses. *American Journal of Psychology*, **47**, 40–64.

WERNER, H. and ZEITZ, K. (1928) Ueber die dynamische Struktur der Bewegung. *Zeitschrift für Psychologie*, **105**, 226–249.

WERTHEIMER, M. (1912) Experimentelle Studien über das Sehen von Bewegung. *Zeitschrift für Psychologie*, **61**, 161–265. Translated in part in T. Shipley (ed.), *Classics in Psychology*. New York: Philosophical Library, 1961.

WERTHEIMER, M. (1923) Untersuchungen zur Lehre von der Gestalt. *Psychologische Forschung*, **4**, 301–350. Translated in part in W. D. Ellis (ed.), *A Sourcebook of Gestalt Psychology*. New York: Humanities Press, 1950.

WESTHEIMER, G. and MITCHELL, D. E. (1969) The sensory stimulus for disjunctive eye movements. *Vision Research*, **9**, 749–755.

WHEATSTONE, C. (1838) On some remarkable and hitherto unobserved phenomena of binocular vision. *Philosophical Transactions*, **128**, 371–394.

WHITE, C. T. (1963) Temporal numerosity and the psychological unit of duration. *Psychological Monographs*, **77**, 12 (Whole No. 575).

WORDEN, F. G. (1966) Attention and auditory physiology. In E. Stellar and J. M. Sprague (eds.), *Progress in Physiological Psychology*, vol. 1. New York: Academic Press.

YOBLICK, D. A. and SALVENDY, G. (1970) Influence of frequency on the estimation of time for auditory, visual, and tactile modalities: The Kappa Effect. *Journal of Experimental Psychology*, **86**, 157–164.

ZAPPAROLI, G. C. and REATTO, L. L. (1969) The apparent movement between visual and acoustic stimulus and the problem of intermodal relations. *Acta Psychologica*, **29**, 256–267.

NAME INDEX

Aarons, L.   17, 175
Adams, P.   85
Aftanas, M.   121
Allan, L. G.   10, 20, 123
Allport, D. A.   147, 148
Allport, F. H.   123
Alpern, M.   99, 131, 132
Ames, A. Jr.   44
Ansbacher, H. L.   145, 146, 147, 148
Anstis, S. M.   58
Arden, G. B.   147
Aubert, H.   31, 32
Axelrod, S.   152

Barlow, H. B.   170
Barr, H. L.   130, 175
Bartlett, F. C.   169
Bartley, S. H.   17, 21, 35, 104
Battersby, W. S.   94, 134
Beazley, L.   144
Bender, M. B.   94, 98, 134
Bill, J. C.   144
Bitterman, M. E.   84
Blakemore, C.   90, 191
Blumenthal, A.   121
Boring, E. G.   2, 7, 21, 149
Bouman, M. A.   27, 29
Boynton, R. M.   131, 147
Braunstein, M. L.   85
Brenner, M. W.   130, 156
Broadbent, D. E.   132
Brooks, L. R.   197
Brosgole, L.   124
Brown, J. F.   31, 32, 59, 60, 61, 67, 144, 145, 180, 181
Brown, R. H.   29, 32, 78
Bruner, J. S.   43, 160, 161
Bullock, T. H.   169

Campbell, F. W.   90, 191
Carran, A. B.   121
Cartwright, D.   31
Chomsky, N.   201
Cohen, J.   143
Cohen, L.   152
Cohen, R. L.   30, 32, 38, 144, 179
Cornsweet, T. N.   161, 173
Corwin, T. R.   147
Crawford, B. H.   131, 135

Deatherage, B. H.   84
DeSilva, H. R.   15, 33, 43, 45, 61, 74, 97, 118, 120, 129, 130, 151, 153, 160, 161, 163, 168, 175, 177, 179
Dimmick, F. L.   55
Djang, S.   105
Dodge, R.   54, 131
Dodwell, P. C.   178, 188
Doherty, R. S.   90, 191
Duncker, K.   20, 120, 122

Ebenholtz, S.   122, 123, 124, 177
Efron, R.   86, 147
Ehrlich, D.   70
Enright, J. T.   17
Ewert, P. H.   17
Eysenck, H. J.   ix
Exner, S.   1, 3, 4, 7, 8, 20

Fehrer, E.   115, 131
Fender, D.   99
Fischer, G. J.   85
Fisichelli, V. R.   85
Fraisse, P.   144
Freedman, S. J.   123
Frisby, J. P.   193

217

Leibowitz, H. W.   30, 32, 38
Lettvin, J. Y.   192
Lewin, K.   189
Lichtenstein, M.   147
Linke, P.   20, 40, 159, 177
London, I. D.   150
Luria, S. M.   135, 136, 169

MacKay, D. M.   71, 174, 200
MacKinnon, G. E.   121
Maffei, L.   191
Mandriota, F. J.   32
Marks, E. S.   144
Marshall, A. J.   146
Mashhour, M.   27, 32, 37
Mates, B.   30, 32
Matin, E.   123, 174
Matin, L.   121, 123, 174
Maturana, H. R.   192
Mayzner, M. S.   185
McCann, J. J.   200
McConnell, R. F.   12, 75
McCulloch, W. S.   192
Miles, W. R.   163
Mintz, D. E.   32
Mitchell, D. E.   47, 173
Moray, N.   148
Morinaga, S.   175
Mosteller, F.   160
Muijen, A. R.   148, 177
Musatti   85

Neff, W. S.   17, 20, 40, 166, 174, 177
Neisser, U.   92, 183
Neuhaus, W.   11, 22, 23, 24, 26, 27, 30, 43, 45, 61, 74, 87, 98, 99, 122, 142, 143, 153, 162, 180, 183
Noguchi, K.   175
Notterman, J. M.   32

O'Connell, D. N.   85, 92
Ogle, K. N.   86, 98
Oldfield, R. C.   43
Orbach, J.   70
Orlansky, J.   45, 46, 47, 110, 181
Osgood, C. E.   15, 59

Pantle, A.   191
Panum, P. L.   85, 98, 99
Pavlov, I.   14
Pearce, D. G.   123
Perkins, D. N.   90, 122, 202
Perky, C. W.   190
Pitts, W.   192
Pola, J.   123, 174
Pollock, W. T.   28, 33, 78
Polyak, S. L.   54
Pomerantz, J. R.   x, 5, 36, 43, 47, 49, 55, 74, 86, 87, 88, 89, 99, 108, 112, 122, 123, 154, 159, 164, 172, 173, 178, 180, 193, 195
Posner, M. I.   47, 150
Postman, L.   160
Pritchard, R. M.   139, 161
Pryluck, C.   187

Raab, D. H.   73, 115, 131
Raskin, L. M.   110, 165
Ratliff, F.   161, 192
Reatto, L. L.   149, 182
Riggs, L. A.   161
Rock, I.   122, 123, 124, 169, 177
Roelofs, C. O.   12, 43, 44, 45, 162
Rogers, R.   149
Rosner, B. S.   99, 132
Ross, P. L.   85
Royce, J. R.   121
Rubin, E.   60
Russel, B.   1, 127

Salvendy, G.   144
Sanders, R. W.   55
Schiller, P. H.   43, 132
Schneider, G. E.   180, 192
Scholz, W.   141, 142, 143, 144, 145
Segal, S. J.   130, 175, 191, 197
Sekuler, R.   125, 191
Sgro, F. J.   183
Shaffer, O.   118
Shapiro, M. B.   84
Sherrick, C. E.   7, 149, 182
Sherrington, C. S.   137
Shipley, T.   13
Shipley, W. C.   13, 182
Siegfried, J. B.   147